Introduction to Oil and Gas Operational Safety

Revision guide for the NEBOSH International Technical Certificate in Oil and Gas Operational Safety

This companion to *Introduction to Oil and Gas Operational Safety* will help you to prepare for the written assessment of the NEBOSH International Technical Certificate in Oil and Gas Operational Safety.

Aligned directly to the NEBOSH syllabus, this revision guide includes learning outcomes and key revision points to help you consolidate your knowledge to enable you to effectively discharge workplace safety and responsibilities.

With reference to the textbook, the guide provides complete syllabus coverage in bite-sized chunks to help you pass the certificate and become an efficient practitioner in the oil and gas industry.

▷ Small, handy size making it ideal for use at home, in the classroom or on the move
▷ Includes revision exercises and answers to check your understanding
▷ Everything you need for productive revision in one handy reference source

Wise Global Training is a company dedicated to providing quality health and safety training in a variety of formats, including eLearning and classroom-based courses as well as webinars.

T0273788

Introduction to Oil and Gas Operational Safety

Revision guide for the NEBOSH International Technical Certificate in Oil and Gas Operational Safety

Wise Global Training

Routledge
Taylor & Francis Group

LONDON AND NEW YORK

First published 2015
by Routledge
2 Park Square, Milton Park, Abingdon, Oxon OX14 4RN

and by Routledge
711 Third Avenue, New York, NY 10017

Routledge is an imprint of the Taylor & Francis Group, an informa business

British Library Cataloguing in Publication Data
A catalogue record for this book is available from the British Library

Library of Congress Cataloging in Publication Data
Introduction to oil and gas operational safety. Revision guide for the NEBOSH international technical certificate in oil and gas operational safety.

pages cm

Includes bibliographical references and index.

1. Petroleum industry and trade–Safety measures–Examinations–Study guides.   2. Petroleum engineering–Safety measures–Examinations–Study guides.   3. Petroleum engineers–Certification.   4. National Examination Board in Occupational Safety and Health–Examinations–Study guides.

TP690.6I643 2015

665.5–dc23

2014016390

ISBN: (pbk) 978-0-415-73078-5
ISBN: (ebk) 978-1-315-84991-1

Typeset in Univers LT by
Servis Filmsetting Ltd, Stockport, Cheshire

Contents

Preface   vii

Illustrations credits   viii

List of abbreviations   x

Revision guide   xv

1   Health, safety and environmental management in context:
    learning from incidents   3

2   Health, safety and environmental management in context:
    hazards inherent in oil and gas   13

3   Health, safety and environmental management in context:
    risk management techniques used in the oil and gas industries   23

4   Health, safety and environmental management in context: an
    organization's documented evidence to provide a convincing
    and valid argument that a system is adequately safe   39

5   Hydrocarbon process safety 1: contract management   47

6   Hydrocarbon process safety 1: process safety management   53

7   Hydrocarbon process safety 1: role and purpose of a
    permit-to-work system   63

8   Hydrocarbon process safety 1: key principles of safe shift
    handover   73

9   Hydrocarbon process safety 1: plant operations and
    maintenance   79

10 Hydrocarbon process safety 1: start-up and shutdown   87

11 Hydrocarbon process safety 2: failure modes   97

12 Hydrocarbon process safety 2: other types of failure   105

13 Hydrocarbon process safety 2: safety critical equipment
    controls   111

14 Hydrocarbon process safety 2: safe containment of
    hydrocarbons   123

15 Hydrocarbon process safety 2: fire hazards, risks and controls   141

16 Hydrocarbon process safety 2: furnace and boiler operations   153

17 Fire protection and emergency response: fire and explosion
    in the oil and gas industries   163

18 Fire protection and emergency response: emergency
    response   177

19 Logistics and transport operations: marine transport   195

20 Logistics and transport operations: land transport   207

Appendix: Answer to Revision Questions   217

Index   281

Preface

NEBOSH International Technical Certificate in Oil and Gas Operational Safety Revision Guide

This companion to the bestselling *Introduction to Oil and Gas Operational Safety* will help you prepare for the written assessments of the NEBOSH International Technical Certificate in Oil and Gas Operational Safety.

It provides complete coverage of the syllabus in bite-sized chunks and will help you learn and memorize the most important areas, with links back to the main *Introduction to Oil and Gas Operational Safety* to help you consolidate your learning.

- ▶ Small and portable, making it ideal for use anywhere: at home, in the classroom or on the move
- ▶ Includes specimen questions and answers
- ▶ Everything you need for productive revision in one handy reference source

Illustrations credits

Figures

3.1   Risk rating matrix. *Source*: Wise Global Training.   26

3.2   HAZOP study flow chart. *Source*: Wise Global Training.   30

3.3   Effective barriers. *Source*: Wise Global Training.   34

7.1   Permit-to-work template. *Source*: Wise Global Training interpretation based on HSE publication HSG250.   68

14.1   Section through external roof storage tank. *Source*: Wise Global Training.   129

14.2   Section through internal roof storage tank. *Source*: Wise Global Training.   130

20.1   Hazard signboard panel. *Source*: Wise Global Training.   208

Tables

3.1   Numerical values of likelihood and consequence. *Source*: Health and Safety Executive.   26

3.2   Hazard checklist. *Source*: New South Wales Department of Planning.   28

13.1   Offshore platform shutdown level response hierarchy. *Source*: Health and Safety Executive.   114

13.2   Onshore process plant shutdown level response hierarchy. *Source*: Health and Safety Executive.   115

15.1   Tools and equipment categorization in zoned areas. *Source*: Health and Safety Executive.   150

15.2   Temperature classification for tools and equipment in zoned areas. *Source*: Health and Safety Executive.   150

19.1   Roles, responsibilities and typical numbers of ship's crew. *Source*: International Labour Organization; www. maritime-transport.net   201

List of abbreviations

ABBI   Above, Below, Behind, Inside
ACGIH   American Conference of Government and Industrial Hygienists
ADR   European Agreement Concerning the International Carriage of Dangerous Goods by Road (Accord Européene Relatif au Transport International des Marchandises Dangereuses par Route)
AFFF   Aqueous Film Forming Foam
ALARP   As Low As Reasonably Practicable
AMA   Advanced Medical Aid
Ar   Argon
ATEX   Atmosphere Explosives. Also, the name commonly given to two European directives for controlling explosive atmospheres
BA   Breathing Apparatus
BLEVE   Boiling Liquid Expanding Vapour Explosion
B-O-P   Blowout Preventer
BOSIET   Basic Offshore Safety Induction and Emergency Training
CCR   Central Control Room
CCTV   Closed Circuit Television
CFD   Computational Fluid Dynamics
CNS   Central Nervous System
CO_2 Carbon dioxide
COMAH   Control of Major Accident Hazards
COSC   Combined Operations Safety Case
CRO   Control Room Operator
CUI   Corrosion Under Insulation
CVCE   Confined Vapour Cloud Explosion
DCR   Design and Construction Regulations

DNA	Deoxyribonucleic Acid
DP	Dynamic Positioning
DPI	Dye Penetration Inspection
DSEAR	Dangerous Substances and Explosive Atmospheres Regulations
DSV	Diving Support Vessel
DTLG	Decommissioning Technical Liaison Group
EBS	Emergency Breathing Systems
ECC	Emergency Command and Control
EER	Evacuation, Escape and Rescue
ELD	Engineering Line Diagram
ELSA	Emergency Life Support Apparatus
ERP	Emergency Response Procedure
ERRV	Emergency Response and Rescue Vessel
ESD	Emergency Shutdown
ESDV	Emergency Shutdown Valve
EU	European Union
F&G	Fire and Gas
FMEA	Failure Mode and Effect Analysis
FMECA	Failure Modes and Effects Criticality Analysis
FPS	Floating Production System
FPSO	Floating Production Storage and Offloading unit
FRC	Fast Rescue Craft
GPA	General Platform Alarm
GPS	Global Positioning System
GRP	Glass Reinforced Plastic
GZ Drum	Gas Zone Drum
H_2S	Hydrogen sulphide
HAZID	Hazard Identification Study
HAZOP	Hazard and Operability Study
HDPE	High Density Polyethylene
HIPPS	High Integrity Pressure Protection System
HLV	Heavy Lift Vessel
HP	High Pressure
HQ	Headquarters
HSE	Health and Safety Executive
HVAC	Heating, Ventilation and Air Conditioning

List of abbreviations

IDLH	Immediately Dangerous to Life or Health
IGC Code	International code for the construction and equipment of ships carrying liquefied gases in bulk
IGG	Inert Gas Generator
ILO	International Labour Office
IMO	International Maritime Organization
KW	Kilowatt
kWm²	Kilowatt per square metre
LEL	Lower Explosive Limit
LNG	Liquefied Natural Gas
LOTO	Lock Out, Tag Out
LP	Low Pressure
LPG	Liquefied Petroleum Gas
LPI	Liquid Penetration Inspection
LSA	Low Specific Activity
MAPP	Major Accident Prevention Policy
MCRC	Maximum Continuous Rating
MEDIVAC	Medical Evacuation
MIG	Manual Inert Gas
MOB	Man Overboard
MODU	Mobile Offshore Drilling Unit
Mpa	Mega Pascal
MPGF	Multipoint Ground Flare
MPI	Magnetic Particle Inspection
MSDS	Material Safety Data Sheet
N	Nitrogen
NDT	Non-Destructive Testing
NORM	Naturally Occurring Radioactive Materials
NSMC	North Sea Medical Centre
NUI	Normally Unmanned Installation
OBM	Oil Based Mud
OIM	Offshore Installation Manager
OSCR	Offshore Installations (Safety Case) Regulations 2005
OSHA	Occupational Safety and Health Administration
P&A	Plugging and Abandonment (of wells)
P&ID	Piping and Instrument Diagram
P&V	Pressure and Vacuum relief valve

PAPA	Prepare to Abandon Platform Alarm
PFEER	Prevention of Fire, Explosion and Emergency Response
PFP	Passive Fire Protection
pH	Power of hydrogen, i.e. a measure of how acidic a substance is
PHA	Process Hazard Analysis
PIG	Pipeline Inspection Gauge
POB	Personnel On Board
PPE	Personal Protective Equipment
ppm	Parts Per Million
PSIC	Protective Systems Isolation Certificate
PSM	Process Safety Management
PSV	Platform Support Vessel
PT	Penetrant Testing
PTW	Permit To Work
RBM	Risk Based Management
RID	European Regulations Concerning the International Carriage of Dangerous Goods by Rail (Règlement Concernant le Transport International Ferroviare des Marchandises Dangereuses par Chemin de Fer)
ROSOV	Remotely Operated Shut-Off Valve
ROV	Remotely Operated Vehicle
RPE	Respiratory Protective Equipment
RT	Radiography Testing
RTU	Remote Terminal Unit
SALM	Single Anchor Leg Mooring
SAR	Search and Rescue
SBM	Single Buoy Mooring
SBM	Synthetic Based Mud
SBV	Standby Vessel
SCC	Stress Corrosion Cracking
SCE	Safety Critical Elements
SIL	Safety Integrity Level
SIMOPS	Simultaneous Operations
SMART	Specific, Measurable, Achievable, Realistic and with Timescales
SMS	Safety Management System

List of abbreviations

SOLAS	Safety of Life at Sea
SOLAs	Specific Off-label Approvals
SPM	Single Point Mooring
SRB	Sulphate Reducing Bacteria
SSCV	Semi-Submersible Crane Vessel
SSIV	Subsea Isolation Valve
STEL	Short Term Exposure Limit
TDS	Total Dissolved Solids
TEMPSC	Totally Enclosed Motor Propelled Survival Craft
TIG	Tungsten Inert Gas
TLP	Tension Leg Platform
TLV	Time Limit Value
TMT	Tube Metal Temperature
TR	Temporary Refuge
Tremcard	Transport Emergency Card
TWA	Time Weighted Average
UEL	Upper Explosive Limit
UKOOA	United Kingdom Offshore Operators Association
UN	United Nations
UNECE	United Nations Economic Commission for Europe
UVCE	Unconfined Vapour Cloud Explosion
VCF	Vapour Cloud Fire
VCT	Vocational Training Certificate
VDU	Visual Display Unit
VESDA	Very Early Smoke Detection Apparatus
VLSSCV	Very Large Semi-Submersible Crane Vessel
WBM	Water Based Mud

Revision guide

When you have completed your study and are ready to take your examination you will sit a 2-hour examination consisting of one paper made up of eleven questions. There will be one main question which will take about 30 minutes to answer, and ten other questions which will take another 90 minutes to answer.

The pass rate is 45 per cent, but if you fail to achieve this mark you will be 'referred', which means you can resit the examination as long as it is within 5 years of the original sitting. There are no limits on how many times you can resit the examination.

Revision guide

Once you have read through all of the material, and completed the course if you are undertaking this as additional support, you should review the content to ensure you fully understand the issues covered. During your revision, make notes of the important points so that you have a summary of the whole subject and can use it as an aide-memoire when preparing for the examination.

We also recommend you purchase a number of past examination papers which are available directly from NEBOSH. NEBOSH also has examiners' reports, which can be useful in understanding how examiners view the way in which candidates generally answered the questions. These reports not only provide an excellent guide on the expected answers to questions but also indicate areas of student misunderstanding.

When NEBOSH sets the examination question papers, it uses what are called 'command words' which dictate the way it expects candidates to answer the questions. Failure to answer the questions in the way the command words indicate can cost marks because they will not have been answered appropriately. Understanding the command words in a question is the key to success in answering it. The command word indicates the nature of the answer and the skills being assessed.

The most frequently used command words include:

▶ **Identify** – This is asking for an answer which selects and names a subject. For example **IDENTIFY** three types of non-destructive testing of welds.

▶ **Give** – This is asking for an answer without an explanation. For example **GIVE** an example of . . . ; **GIVE** the meaning of . . .

▶ **Outline** – This is asking for an answer which gives the principal features or different parts of a subject or issue. An exhaustive description is not required. What is sought is a brief summary of the major aspects of whatever is stated in the question.

▶ **Describe** – This is asking for a detailed written account of the distinctive features of a subject. The account should be factual without any attempt to explain. A further definition of **DESCRIBE** as a command word is a picture in word form.

▶ **Explain** – This is asking for the reasoning behind, or an account of, a subject. The command word is testing the candidate's ability to know or understand why or how something happens.

Element 1

CHAPTERS 1–4

Sub-element 1.1:
Learning from incidents

Learning outcome

Explain the purpose of, and procedures for, investigating incidents and how the lessons learnt can be used to improve health and safety in the oil and gas industries.

🔑 Key revision points

Investigating incidents and effective identification of the root causes and making recommendations for improvements

☐

Importance of learning lessons from major incidents, management, culture and technical failures that may lead to such incidents

☐

Definitions

▶ **Accident** – an unplanned event that results in damage, loss or harm
▶ **Hazard** – the potential of something to cause harm
▶ **Risk** – the likelihood of something to cause harm
▶ **Residual risk** – remaining risks after controls have been applied
▶ **Near miss** – any unplanned incident, accident or emergency which did not result in an injury but which could have done
▶ **Dangerous occurrence** – a near miss that could have led to serious injury or loss of life
▶ **Damage only** – an event which caused damage but has not involved injury to anyone but which could have done
▶ **Outcome** – the effect of an unplanned, uncontrolled event
▶ **Minor injury** – an injury which does not involve time off work
▶ **Significant injury** – an injury which is not major but which dresults in the injured person being away from work or being unable to do their full range of normal duties
▶ **Major injury** – an injury which can be regarded as a serious threat to a person's health and/or well-being

Investigating incidents and effective identification of the root causes and making recommendations for improvements

Why accidents and incidents should be investigated

▶ To determine their cause
▶ So that information forthcoming from the investigation can be used to avoid it happening again

Legal reasons for investigating accidents and incidents

▶ To demonstrate that the company is meeting its legal requirements.

▶ Employers should be able to make available information regarding the circumstances appertaining to the accident in case those involved in the accident decide to take legal action.

▶ If needs be, a company can demonstrate to the courts their commitment and positive attitude to health and safety.

Financial reasons for investigating accidents and incidents

▶ Information forthcoming from an accident investigation provided to an insurance company may well assist in the event of a claim.

▶ The outcome of an investigation could prevent a recurrence with potential catastrophic results.

▶ The company can avoid business losses if they take heed of the outcome of an investigation by preventing further accidents or incidents.

▶ Other costs saved might include the cost of legal action which may be taken against the company; increased insurance premiums; loss of business due to a bad reputation resulting in lost orders.

Other reasons for investigating accidents and incidents

▶ Establish how and why an accident/incident happened

▶ Establish exactly what working practices and procedures are

▶ Discover how exposure to conditions (e.g. noise, cold, heat), or substances (e.g. chemicals, radiation, gases) may affect the health of employees

▶ Expose weaknesses or faults in production systems whereby a certain scenario of events will expose these weaknesses or faults

▶ Lessons learned in one department of an organization can be shared with other departments

Benefits from investigating accidents and incidents

▶ The outcomes of an investigation can result in the company putting measures in place to prevent the recurrence of similar accidents or incidents in the future.

▷ The development of a health and safety culture within the company: following an investigation, any measures which are put in place as a result of the findings will be more readily accepted by the workforce, especially if they were involved in the decision-making process.

▷ Managerial skills will be developed during any investigation and these can be used in other departments within the organization.

The investigation team, typical make-up

▷ Operations team leader
▷ In the case of an offshore installation, a field or platform safety officer
▷ In the case of an onshore installation, a senior onshore manager
▷ Safety representatives
▷ Area authorities (the person responsible for the area where the incident took place)
▷ Specialist inspectors
▷ If it's a drilling rig, a tool pusher

Knowledge required and training for the investigating team

▷ Knowing their own roles and responsibilities
▷ Knowing which events need to be reported
▷ Knowing accident book regulations and requirements and how to use it as a source of historical information
▷ Knowing which documents and forms relevant to the investigation – internal and external – need to be completed
▷ Knowing the importance of reporting accidents/incidents/dangerous occurrences/near misses for legal, investigative and monitoring reasons
▷ Knowing how the dissemination of information will be conducted and to whom

Four stage process of investigating accidents/incidents

Step one – gathering the information

Step two – analysing the information

Step three – identifying the required risk control measures

Step four – formulation of the action plan and its implementation

Investigating accidents/incidents – observational techniques

Good observation skills and techniques should include:

▶ Knowledge of the workplace and procedures

▶ Keeping a systematic record of observations

▶ Taking time to observe the whole scene

▶ Being alert to possible changes to the accident scene by those who may have a motive to correct unsafe practices

▶ Using the ABBI technique: look Above, Below, Behind, Inside

▶ Being inquisitive and questioning employees to determine risks – their views can be a valuable source of insight

▶ Using all senses including smell, sight, touch and hearing

▶ Having an open mind and looking for solutions

▶ Identifying, recording and feeding back good performance as well as bad

▶ Using an interviewing style which does not reflect a blame culture

▶ Asking questions in a way which does not make the interviewee feel intimidated or uncomfortable

▶ Conducting the interview in familiar surroundings as this will be less intimidating

▶ Encouraging co-operation by allowing witnesses to speak openly in their own words without using technical jargon

▶ Promoting a positive attitude to finding the reasons for the incident to prevent a recurrence in the future rather than apportioning blame for the present one

▶ Interviewing witnesses separately and in private to prevent them from influencing each other's accounts

7

▶ Providing a summary of what the witness said in order that they can ensure that everything has been understood correctly and that the interviewer has not misinterpreted the account

Investigating accidents/incidents – relevant records and sources of information

▶ Victim statements
▶ Witness statements
▶ Plans and diagrams
▶ CCTV coverage
▶ Process drawings, sketches, measurements, photographs
▶ Check sheets, permits-to-work records, method statements
▶ Details of the environmental conditions at the time
▶ Written instructions, procedures and risk assessments which should have been in operation and followed
▶ Previous accident records
▶ Information from health and safety meetings
▶ Technical information/guidance/toolbox talk sheets
▶ Manufacturers' instructions
▶ Risk assessments
▶ Training records
▶ Logs
▶ Instrument readouts and records
▶ Opinions, experiences, observations

Investigating accidents/incidents – analysing the information

▶ Be objective and unbiased.
▶ Identify the sequence of events and conditions that led up to the event.
▶ Identify the immediate causes.
▶ Identify underlying causes, i.e. actions in the past that have allowed or caused undetected unsafe conditions/practices.
▶ Identify root causes, i.e. organizational and management health and safety arrangements – supervision, monitoring, training, resources allocated to health and safety, etc.

Investigating accidents/incidents – analysing the causes

▶ **Immediate causes** are generally unsafe acts and/or conditions.
▶ **Underlying causes** are generally procedural failures.
▶ **Root causes** are generally management system failures.

Investigating accidents/incidents – hierarchy of risk control

▶ Eliminate the risk altogether.
▶ Replace the risk with something safer.
▶ Apply engineering controls such as cut out devices, guards, etc.
▶ Apply administrative controls such as safe working practices.
▶ Use Personal Protective Equipment (PPE), but only as a last resort or in conjunction with other controls.

Investigating accidents/incidents – SMART objectives

▶ **S**pecific
▶ **M**easurable
▶ **A**chievable
▶ **R**ealistic and with
▶ **T**imescales

Learning lessons from major incidents

Learning lessons locally

▶ For the management of the organization – they will need a report which details what went wrong; the systems and procedure failures which were involved; how serious the incident could have been
▶ How to avoid future incidents
▶ For regulatory bodies, safety records, etc. – technical details of the investigation as well as its findings
▶ For operators of the systems and procedures involved in the incident

9

▶ For incident investigators generally. This group of people have an ongoing need to broaden their knowledge and understanding of how things go wrong and how to encompass this knowledge in future investigations.

Learning lessons more widely

▶ Lessons learned within one organization can be disseminated widely throughout other organizations by the publication of information in trade or specialized journals or publications, or through internet websites.

Sub-element 1.1: Learning from incidents

Learning outcome

Explain the purpose of, and procedures for, investigating incidents and how the lessons learnt can be used to improve health and safety in the oil and gas industries.

✎ Revision exercise

Write your answers to the questions below on a separate sheet of paper without referring to the information in this book in the first instance. Once you have answered all of the questions, you can refer back to the revision guide to compare your answers, and this will give you an indication of how much knowledge you have been able to absorb or whether you need to revise this section further.

Q1 **Outline FIVE** reasons for investigating accidents and incidents. Include in your answer at least one legal reason and two financial reasons.

Q2 **Identify** the kind of persons you would expect to be involved in a team set up to investigate an accident/incident.

Q3 **Identify** the **FOUR** stages in the process of investigating accidents and incidents.

Q4 **Identify SIX** observational skills or techniques used in accident/incident investigations.

Q5 **Give SIX** sources of information relating to accident/incident investigations.

Q6 In relation to causes of accidents, **explain** what is meant by:
  (a) immediate causes
  (b) underlying causes
  (c) root causes

Q7 **Give** the hierarchy of risk control.

Q8 Lessons learnt from major incidents can be disseminated both locally and more widely. **Outline** who might benefit from lessons learnt:
  (a) locally
  (b) more widely

CHAPTER 2

Sub-element 1.2:
Hazards inherent in oil and gas

Learning outcome

Explain the hazards inherent in oil and gas arising from the extraction, storage and processing of raw materials and products.

🔑 Key revision points

Meaning and relevance of various words and phrases associated with hazards inherent in oil and gas

Properties and hazards of various gases associated with the oil and gas industry

Properties and hazards of associated products and their control measures

Meaning and relevance of various words and phrases associated with hazards inherent in oil and gas

Flash point

The **flash point** of a volatile liquid is the lowest temperature at which it can vaporize to form an ignitable mixture when mixed with air.

Vapour density

Vapour density is the measurement of how dense a vapour is in comparison with air.

Vapour pressure

The process of evaporation involves the molecules on the surface of a liquid. When the energy within these molecules is sufficient for those molecules to escape, they do so in the form of a vapour. This is known as **vapour pressure**.

Flammability

Flammable – This describes a product which is easily ignitable and capable of burning rapidly. Note that the word inflammable has the same meaning as flammable. In the UK a flammable liquid is defined as a liquid that has a flash point of between 21°C and 55°C. However, in the USA there is a precise definition of flammable liquid as one with a flash point below 100°F (37.8°C).

Highly flammable – This describes a product which has a flash point below 21°C but which is not defined as extremely flammable.

Extremely flammable – This describes a product which has a flash point lower than 0°C and a boiling point of 35°C or lower.

Fire triangle – For a fire to start, there are three elements which have to be present. These are:

1   A source of fuel
2   A source of ignition
3   Oxygen

14

These three elements are known as the **fire triangle**.

Flammable range – The **lower flammable limit** is the lowest concentration of a gas or vapour in air which is capable of being ignited. The **upper flammable limit** is the highest concentration of a gas or vapour in air which is capable of being ignited. The percentage of flammable vapour which falls between these two parameters is known as the **flammable range**.

Toxicity

▷ **Acute toxicity** is a term which describes the effect a substance has had on a person after either a single exposure or from several exposures within a short space of time (e.g. 24 hours or less).
▷ **Chronic toxicity** is a term which describes the effects a substance has had on a person after many exposures over a longer period of time (e.g. months or years).

Skin irritant – The Occupational Safety and Health Administration (OSHA) describes skin irritant as 'a chemical, which is not corrosive, but which causes a reversible inflammatory effect on living tissue by chemical action at the site of contact'.

Carcinogenic properties – A carcinogen is defined as any substance that can cause, or aggravate, cancer. They fall into two groups:

▷ **Genotoxic carcinogens** are those which react with DNA directly or with macromolecules which then react with DNA.
▷ **Non-genotoxic carcinogens** do not react directly with DNA although they do cause cancer in other ways.

Properties and hazards of various gases associated with the oil and gas industry

Hydrogen is a gas which is difficult to detect as it is odourless and colourless. It is lighter than air (a density of 0.07 when compared with air) and so will rise when released.

▶ **Hazards** associated with hydrogen include: it is a highly flammable gas when it is mixed with air (flammable range 4–75 per cent); it burns with an invisible flame.

Hydrogen sulphide (H_2S) is produced from decaying vegetation and marine micro-organisms. It can be released as it comes to the surface with drilling shale.

▶ **Hazards** associated with hydrogen sulphide include: it is a toxic, corrosive and flammable gas; it has a density of 1.39 when compared with air; tends to drift in low lying areas such as pits, cellars, drains; and is difficult to disperse.

Methane is an odourless, colourless gas which exists naturally in the substrate. It is lighter than air with a density of 0.717 compared with air.

▶ **Hazards** associated with methane include: it is a flammable gas and, when mixed with air in concentrations between 5 and 15 per cent, is explosive; it can cause asphyxiation if the concentration is high enough.

Liquefied Petroleum Gas (LPG) is a mixture of hydrocarbon gases which are highly flammable. It is an odourless, colourless gas which has a density of 2.0 when compared with air and tends to drift in low lying areas such as pits, cellars, drains, etc. As such, it is difficult to disperse.

Liquefied petroleum gas expands at a rate of 250:1 at atmospheric pressure when it changes from a liquid to a gas.

▶ **Hazards** associated with LPG include: it can cause a massive vapour cloud from a relatively small amount of liquid when that liquid is released into the air, with the potential to cause a Boiling Liquid Expanding Vapour Explosion (BLEVE); it can cause asphyxiation; it can cause cold burns to the skin on contact; it can cause brittle fracture to carbon steel on contact; it can cause environmental damage.

Liquefied Natural Gas (LNG) is a colourless, odourless, highly flammable natural gas which is made up of methane (85–95 per cent), ethane, propane and butane. It is non-corrosive and non-toxic.

▶ **Hazards** include: it can cause asphyxiation; it can cause cold burns to the skin; it can cause brittle fracture to carbon steel on contact.

Nitrogen is a colourless, odourless, non-flammable gas which is often used as a blanket gas in storage tanks and for purging equipment and processes of oxygen and hydrocarbons, thus eliminating the hazards of fire and explosion.

▶ **Hazards** associated with nitrogen include potential asphyxiation when it is used in confined spaces to displace oxygen.

Oxygen is an odourless, colourless gas which is present in the atmosphere.

▶ **Hazards** associated with oxygen include the potential to cause asphyxiation. This is because the body is stimulated to breathe by the level of carbon dioxide (CO_2) in the air and should a situation arise where oxygen is released into an area displacing the carbon dioxide, then the stimulus to breathe could cease, causing death by asphyxiation. It is also one element of the 'fire triangle', i.e. it allows a fire to burn. It can also cause rusting.

Properties and hazards of associated products and their control measures

Anti-foaming agents and anti-wetting agents

Anti-foaming agents are used to prevent foam forming or to break down foam that has already been created in a process liquid during any production process.

Anti-wetting agents are coatings which are applied to surfaces of vulnerable components which are subject to moisture and subsequent corrosive activities.

▶ **Control measures** – Although both anti-foaming agents and anti-wetting agents are generally non-hazardous, it is advisable to wash the area of any skin contact with soap and water; information provided by the product's Material Safety Data Sheet (MSDS) should always be consulted when they are used.

17

Micro-biocides are used to protect against the harmful effects of bacteria, e.g. legionella, which can proliferate in air conditioning systems and humidifiers. Micro-biocides are classed as irritants to skin and eyes on contact as well as being toxic if ingested.

▶ **Control measures** – Information provided by the product's Material Safety Data Sheet (MSDS) should always be consulted when they are used.

Corrosion preventatives – Some corrosion preventatives come in the form of a water displacing film which acts by spreading across the surface of metals, displacing water from cracks and crevices and forming a barrier to corrosive activity. Other types of preventatives are applied to metals which then dry to a hard resin or waxy film, thus forming a barrier to corrosive activity.

▶ **Control measures** – Information provided by the product's Material Safety Data Sheet (MSDS) should always be consulted when they are used.

Refrigerants are liquefied gases under pressure which are used to reduce temperature in certain situations.

▶ **Hazards** associated with refrigerants include: injury from components or material ejected by the high pressure escape; frostbite; asphyxiation; possible explosion; possible secondary toxic gases if refrigerant gases burn; as refrigerant gases are heavier than air they will tend to drift in low lying areas such as pits, cellars, drains and are difficult to disperse.

▶ **Control measures** – Have procedures in place to deal with any unexpected release; never work in confined spaces where there is a risk that refrigerants may be released; provide ventilation; immediately remove anyone who has been exposed to refrigerant gases.

Water is used for cooling and dilution in process operations as well as for fighting fires, cleaning and within air conditioning systems. The hazards associated with water include: Legionella from air conditioning systems; leptospira from fresh water rivers or lakes; corrosion of steel components; electrostatic charge; possible failure of components from high pressure within pipework; possible failure of components from frozen water.

▶ **Control measures** for frozen water include lagging pipes; fitting steam tracer lines; draining unused components.

Sea water contains living organisms which can proliferate and cause blockages.

▶ **Control measures** for sea water include implementing a regular maintenance programme; applying additives to kill any living organisms; fitting a dry riser.

Steam is used to power turbines and generate electricity, as well as serving as a source of heat and/or energy to assist with many other operations and processes. It can also be used to protect systems from the risk of freezing (tracer lines) and to serve as a heating system for areas where personnel are housed.

▶ **Hazards** associated with steam include: the potential to cause thermal shock; high pressure steam can cause failure of components; it can cause burns to skin.

Mercaptans are used to help detect the presence of natural gas by giving it an odour.

▶ **Hazards** include the fact that some mercaptans are harmful. For example, methyl mercaptan is harmful if inhaled; it is a respiratory irritant – chronic exposure may cause lung damage; it is a skin and eye irritant; it can depress the central nervous system; it has a flashpoint of −18°C.

Drilling muds – When drilling operations are in progress, mud is pumped from the mud pits through the 'drill string' where it is sprayed onto the drill bit. This allows for the cooling and cleaning of the drill bit throughout its operation. It also assists in suspending and the eventual removal of cuttings from the well, transmitting hydraulic energy to tools and bit, controlling formation pressure, sealing permeable formations, minimizing formation damage, maintaining wellbore stability, ensuring adequate formation evaluation, controlling corrosion and facilitating cementation on completion.

▶ **Hazards** associated with drilling muds include the fact that when the muds come to the surface there may be natural gases or

19

other flammable materials which have combined with the mud during the drilling operation. These have the ability to be released from the mud anywhere within the system where the mud is flowing back to the pit. As a result of this there is risk of fire or explosion should these gases be exposed to a source of ignition.

▷ **Control measures** include applying safe working procedures; installing monitoring sensors; using equipment and wiring which has been certified as explosion proof.

Water based drilling mud is a combination of clay and/or other additives such as bentonite and potassium blended together with water. These give the drilling mud various characteristics such as viscosity control, shale stability, enhancing the drilling rate of penetration and cooling and lubricating the drilling equipment.

Oil based drilling muds have, as their name suggests, oil – usually diesel oil – as their base fluid. The advantages oil-based muds bring to the drilling process include: increasing lubrication of the drill shaft, enhancing shale inhibition, adding greater cleaning ability, allowing for higher working temperatures to be used without adverse effects.

▷ **Hazards** include being toxic and having the potential to cause environmental damage.

Synthetic-based drilling muds have the properties of oil based muds but have the advantage of being less toxic as their base is made from synthetic oil.

Low specific activity sludges – The formations of rock and shale, which contain oil and gas deposits, also contain Naturally Occurring Radioactive Materials (NORM). These include uranium, thorium, radium and lead-210 which dissolve in the brine, and these eventually separate out and form wastes at the surface. The process of extraction exposes the environment and humans to the radioactive elements in the sludges. As such, they are classified as hazardous. They can be found on various pieces of equipment and locations including on the drill string, inside vessels (demister pads), inside filters, in coalescars (coarse filter/emulsifier) and in coolers where tubes might be coated with sludge.

▶ **Hazards** from the sludges, which are a mixture of liquid and suspended material, include: the inhalation and ingestion of radionuclides, especially dust and fumes; skin irritant (possibly causing dermatitis); inhalation (fumes or dust from dried sludges); direct radiation potential causing carcinogenic problems; environmental pollution.

▶ **Control measures** include: the provision of ventilation equipment to control dusts and fumes; the use of wet methods of working and good housekeeping to reduce the amount of dust in the atmosphere; having equipment in place to collect sludge instead of using manual means; diluting sludge with water; the use of permit-to-work systems; the provision of training and awareness programmes; the provision of a health surveillance programme to monitor the health of employees; the use of Respiratory Protective Equipment (RPE) specifically chosen to protect against exposure to airborne radioactivity.

Sub-element 1.2: Hazards inherent in oil and gas

Learning outcome

Explain the hazards inherent in oil and gas arising from the extraction, storage and processing of raw materials and products.

Revision exercise

Write your answers to the questions below on a separate sheet of paper without referring to the information in this book in the first instance. Once you have answered all of the questions, you can refer back to the revision guide to compare your answers and this will give you an indication of how much knowledge you have been able to absorb or whether you need to revise this section further.

Q1 **Outline** the features of the following:
  (a) flash point

(b) vapour density

(c) vapour pressure

Q2 **Explain** the difference between 'flammable', 'highly flammable' and 'extremely flammable'.

Q3 **Explain** what the 'lower flammable limit' and 'upper flammable limit' of a product is.

Q4 Toxicity can be 'acute' or chronic'. **Explain** the difference between the two terms.

Q5 Some substances are described as 'skin irritants'. **Explain** what a 'skin irritant' is.

Q6 **Explain** what is meant when a substance is said to have 'carcinogenic properties'.

Q7 **Outline** hazards associated with:

(a) hydrogen gas

(b) hydrogen sulphide

(c) methane

(d) Liquefied Petroleum Gas (LPG)

(e) Liquefied Natural Gas (LNG)

(f) nitrogen

(g) oxygen

(h) micro-biocides

(i) refrigerants

(j) steam

(k) mercaptans

Q8 **Outline** hazards and control measures associated with:

(a) drilling muds

(b) low specific activity sludges

Sub-element 1.3:

Risk management techniques used in the oil and gas industries

Learning outcome

Outline the risk management techniques used in the oil and gas industries.

🔑 Key revision points

The purpose and uses of risk assessment techniques, qualitative and quantitative techniques ☐

How risk management tools are applied in process safety risk identification and assessment, application in project phases from concept, design, start up, the concept of ALARP and the management of major incident risks ☐

Industry-related process safety standards, inherent safe and risk based design concepts, engineering codes and good practice ☐

The concept of hazard realization ☐

The concept of risk control using barrier models ☐

Use of modelling such as thermal radiation output, blast zones for risk identification ☐

The purpose and uses of risk assessment techniques, qualitative and quantitative techniques

There is a responsibility within the oil and gas industry to identify those risks with the potential to cause fire, explosion, environmental contamination and injury to personnel and put in place control measures to reduce them to a level that is as low as is reasonably practicable. This requires the implementation of various risk assessment techniques, including both qualitative and quantitative techniques.

What a risk assessment is and its purpose

In order to put in place risk control measures it is important to identify those risks, the first step of which is to perform a risk assessment. This allows those risks which are relevant to be identified and be given appropriate consideration as to their control. There are a number of techniques available when assessing risks, including the 5-step approach; qualitative assessment techniques; semi-quantitative assessment techniques; and quantitative assessment techniques.

The 5-step approach to risk assessment

Step 1   Identify the hazards
Step 2   Decide who might be harmed and how
Step 3   Evaluate the risks and decide on precautions
Step 4   Record the findings and implement them
Step 5   Review the assessment on a regular basis and update if necessary

Hierarchy of risk control is a means of prioritizing measures to reduce risk with those control measures at the top of the list being given the highest priority.

- Elimination
- Substitution
- Engineering controls
- Administrative controls
- Personal Protective Equipment (PPE)

Qualitative risk assessment is based on the conclusions reached by the assessor using his/her expert knowledge and experience to judge whether current risk control measures are effective and adequate in order to ensure they reduce the risk to a level which is as low as is reasonably practicable, or if more measures need to be applied. When making a qualitative judgement on the severity of a risk, two parameters are taken into consideration. These are the likelihood of an event occurring and the consequences or severity if the event does occur.

Semi-quantitative risk assessment involves applying a numerical value to the likelihood of a particular event occurring and the degree of severity if it should occur. An example of the kind of rating used, where measures ranging from 1 to 5 are applied, is given in Table 3.1.

25

Table 3.1 – Numerical values applied to levels of likelihood and consequence

Likelihood can be defined as	Severity can be defined as
5 Very likely	5 Catastrophic
4 Likely	4 Major
3 Fairly likely	3 Moderate
2 Unlikely	2 Minor
1 Very unlikely	1 Insignificant

Source: Adapted from www.hse.gov.uk/quarries/education/overheads/topic5.doc by David Mercer.

When judging the risk of a particular activity, the risk assessor or risk assessment team agree the likelihood rating, e.g. 3, agree the consequence (severity) rating, e.g. 4, then multiply the likelihood (3) by the consequence (4) to get a rating of 3 × 4 = 12 (tolerable). This can be seen in the matrix shown in Figure 3.1 below. This form of semi-quantitative risk rating system gives an overall numerical value on the risk being evaluated. That numerical value can then be used to prioritize the actions required, as shown in the grading on the right of the matrix.

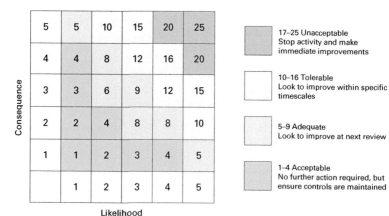

Figure 3.1   Risk rating matrix

Source: Wise Global Training.

Quantitative risk assessment involves using special quantitative tools and techniques in order to identify hazards and to give an estimate of the severity of the consequences and the likelihood of those hazards being realized. The quantitative risk assessments will result in the provision of numerical estimates of the risks, and these can then be evaluated when compared with known numerical risk criteria. Historical data analysis is the basis for many quantitative risk assessments. Frequencies are simply calculated by combining accident experience and population exposure, typically measured in terms of installation-years:

$$\frac{\text{Number of events}}{\text{Number of installations} \times \text{Years of exposure}}$$

One example of a source of historical data which can be used as the basis for quantitative risk assessments is the Worldwide Offshore Accident Databank (WOAD).

How risk management tools are applied in process safety risk identification and assessment; application in project phases from concept, design and start-up; the concept of ALARP; and the management of major incident risks

When a project is in the design stage, some risks can be 'designed out', as can some of the hazards, using modelling as a tool. Examples of modelling techniques include HAZID (Hazard Identification Studies); HAZOP (Hazard and Operability Studies); FMECA (Failure Modes and Effective Critical Analysis)/FMEA (Failure Modes and Effects Analysis).

HAZID (Hazard Identification Study) is a tool for identifying hazards. It is normally a qualitative risk assessment and is judgement based. It is usually undertaken by a team of people who will be selected because of their particular knowledge, experience or expertise. The reasons for identifying hazards are twofold.

27

3 Risk management techniques

1 To compile a list of hazards. This can then be evaluated using further risk assessment techniques. This may be described as 'failure case selection'.

2 To conduct a qualitative evaluation of how significant the hazards are and how to reduce the risks associated with them. This may be described as 'hazard assessment'.

The following are examples of the keywords and some of their associated hazards used as prompts during the study process.

Table 3.2 – Hazard checklist

Key words	Hazards
Fire	Blowout which has ignited
	Process leak which has ignited
	Product storage leak which has ignited
Loss of breathable atmosphere	Ingress of smoke
	Asphyxiation
Toxic gas release	Ingress of toxic gas
	Asphyxiation
LPG/LNG release/leak	Explosion from contact with a source of ignition
	Hydrate formation on valves
	Cold burns/frostbite
	Brittle fracture of steel component(s)
Hydrocarbon release/leak	Explosion from contact with a source of ignition
Collision/crash	Helicopter crash
	Vessel colliding with rig
Structure failure	Crane collapse
	Rig leg collapse

Source: Adapted from New South Wales Department of Planning (2011) *Hazardous Industry Planning Advisory Paper No. 8*. HAZOP Guidelines, Sydney, State of New South Wales ISBN 978 0 73475 872 9 available at: http://www.planning.nsw.gov.au/Portals/0/HIPAP%208%20Final%202011.pdf

HAZOP (Hazard and Operability Study) is a tool which is used to systematically examine every part of a process or operation in order to find out how deviations from the normally intended operation of a

28

process can happen and if further control measures are required to prevent the hazards from happening.

Every part of the installation is systematically examined by a team which comprises experts with a wide range of skills and experience relevant to the installation. This allows the questions to be asked to explore every possible way the operation could deviate from the normal intended operation of the process, and thus test its integrity. The questions asked are set around a set of guide words.

The Hazard and Operability Study (HAZOP) team should consist of a group of people who, between them, have expert knowledge in every area of the process plant and its operations. Typically they are a group of between five to eight people in the fields of management and engineering.

An example of a Hazard and Operability Study (HAZOP) team who have been assembled to consider the design of a new chemical plant could comprise the following people:

- Chairperson
- Design engineer
- Process engineer
- Electrical engineer
- Instrument engineer
- Operations engineer

FMECA (Failure Modes and Effects and Criticality Analysis) is a method of systematically identifying the failure modes of an electrical or mechanical system. One or two people examine each component of the system in turn and evaluate the effects and the degree of importance if that component should fail. The examination uses a document that contains a systematic list of all of the components and usually includes:

- The name of the component
- The function of the component
- The possible types of failure
- The causes of each type of failure
- How each failure is detected
- The effects of each failure on the primary system

29

3 Risk management techniques

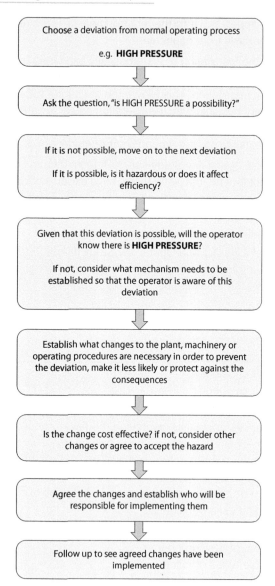

Figure 3.2   HAZOP study flow chart

Source: Wise Global Training.

- The effects of failure on other components
- Necessary actions to prevent each failure or what actions are necessary to repair each failure
- The degree of criticality

As Low As Reasonably Practicable (ALARP)

What this means is that employers should adopt appropriate safety measures unless the cost (in terms of money, time or trouble) is grossly disproportionate to the risk reduction. Once all such measures have been adopted, the risks are said to be 'as low as reasonably practicable'.

Management of major incident risks

Management of major incident risks should take a hierarchical approach:

1 Elimination and minimization of hazards (designing safety into process and systems)
2 Prevention (the reduction of the likelihood of a major incident)
3 Detection (the warning and alarm systems transmitted to the control area)
4 Control (the limitation of the scale, the intensity and/or the duration of an incident)
5 Mitigation of consequences (the protection from effects of an incident)

Following the report on the Buncefield incident, which occurred in December 2005, the principal recommendations for the management of major incident risks were as follows:

- To review emergency arrangements to cover all reasonably foreseeable emergency situations which may result from credible major incidents
- To ensure that guidance, which is related to existing emergency plans, is reviewed by an external, independent authority
- To ensure training is given to all relevant staff in order that they become competent in the implementation of the emergency plan

31

▶ To ensure that there is an adequate level of trained staff available at all times

▶ To ensure that the control centre used for an emergency situation is appropriately sited and adequately protected

▶ To ensure all critical emergency response resources are identified and contingency plans exist in case any of them fail

▶ To ensure that adequate arrangements with all external emergency services have been made

▶ To ensure that there is regular communication between the operator of the installation and any external agencies that may be affected by it

Industry related process safety standards, inherent safe and risk based design concepts, engineering codes and good practice

Industry related process standards

The oil and gas industry is a multi-national operation and is governed by both national and international health and safety regulations and codes of practice which are developed and enforced by government departments and other authorities throughout the world (e.g. the Occupational Safety and Health Administration (OSHA), the Health and Safety Executive (HSE), etc.). Governments and enforcing authorities tend to work in conjunction with the oil and gas industry in order to develop these codes of practice and legislation, which emerge from the specialized knowledge and experience which these industries have in managing their risks.

Inherent safe and risk based design concepts

One of the main elements of developing inherently safe processes is to reduce the complexity of the plant at the design stage and simplify the operation process. Designers can use a hierarchical structure, with hazard avoidance being the priority followed by the control of any risks remaining. The main principles for achieving an inherent safe design are as follows.

32

▷ Minimizing the amount of hazardous material present at any one time

▷ Substituting hazardous materials with less hazardous materials

▷ Moderating the effect a material or process might have (reduce temperature or pressure)

▷ Simplifying the design by designing out problems rather than adding features to deal with problems

▷ Designing in tolerance levels to cope with faults or deviations

▷ Limiting the effects of any adverse event, e.g. by installing bunds around storage tanks

▷ Allowing for human error by designing in failsafe features such as valves which fail to a SHUT position

The concept of hazard realization

Hazard realization

Hazard realization is when a system of hazard controls breaks down or fails, which in turn causes a hazardous event to occur.

The concept of risk control using barrier models

Barrier modelling

The definition of a barrier is something which is placed between a person and a hazard to prevent that person from being harmed. Examples include barrier cream, which is applied to protect hands from harmful substances, and a machine guard. Barriers can also be intangible things such as knowledge and training.

Control measures in place to mitigate potential exposure

The model shown in Figure 3.3 illustrates how risk can be minimized by putting in several barriers. Although every barrier can be breached in a number of ways, each of them reduces the exposure to some

33

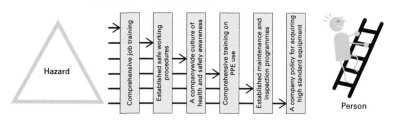

Figure 3.3 Effective barriers

Source: Wise Global Training.

extent. In order for the hazard to be realized it would be necessary for all of the barriers to be breached simultaneously.

Examples of good barriers are:

▷ Good design and specifications
▷ Good processes and procedures
▷ Robust inspection and maintenance techniques
▷ Adequately trained and competent personnel

Use of modelling such as thermal radiation output, blast zones for risk identification

There are a number of modelling tools available to help identify the risk of fire and explosion. The risks involved depend on the type of fire and their individual characteristics, so it is useful to look at all relevant fires and the modelling techniques associated with them in order to help in risk identification and subsequent risk reduction.

Pool fire

A pool fire is a fire burning above a horizontal and stable pool of vaporizing hydrocarbon fuel. If the fuel is not stable, it is known as a running fire. As the pool increases in area, the proportion of fuel burning off increases until it eventually matches the rate of input. At this time the pool should remain constant in size.

Jet fire

A jet fire is a flame which is being fed by hydrocarbons continuously being released with significant momentum in a particular direction. Assessment of the hazards of jet fires is made by analysing the length of the jet flame relative to its distance from equipment, buildings, people, etc.

Flash fire and fire ball

When a dense cloud of vapour is formed by the release of flammable gases or liquids and this meets a source of ignition, a Vapour Cloud Fire (VCF) may result. This is also known as a flash fire or fire ball. The duration, height and diameter as well as the amount of uplift are all characteristics of flash fires and fireballs which can be modelled using a formula based on the mass of fuel released.

Explosion hazard assessment modelling

Empirical models (knowledge acquired by means of observation or experimentation) employ a simplified version of the physics of an explosion and are useful for quickly calculating the order of magnitude of explosions as well as for screening scenarios which then require further investigation with more sophisticated tools.

Phenomenological models (study of structures) are more complicated than empirical models. They are models which 'fit' the experimental data and are used to represent the scenario geometry in simple terms.

Computational Fluid Dynamics (CFD) models fall into two groups – simple models and advanced models. Advanced models provide a more complete description of the physical and chemical processes involved in explosions, including an improved representation of the geometry and accuracy of the numerical schemes.

Explosion consequence assessment modelling

This type of modelling looks at the consequences of explosions. For example, injury to people from an explosion can be as a direct consequence of the blast wave (e.g. rupture of the ear drum) or indirectly (e.g. from flying debris). Equipment and structures can be damaged by the effects of loading (affecting walls and large objects) or drag loading (affecting steelwork or pipework which are constructed of narrow cross-sections) or a combination of both.

Sub-element 1.3: Risk management techniques used in the oil and gas industries

Learning outcome

Outline the risk management techniques used in the oil and gas industries.

✎ Revision exercise

Write your answers to the questions below on a separate sheet of paper without referring to the information in this book in the first instance. Once you have answered all of the questions, you can refer back to the revision guide to compare your answers and this will give you an indication of how much knowledge you have been able to absorb or whether you need to revise this section further.

Q1 **Give** the five steps in a risk assessment process.
Q2 **Give** the hierarchy of risk control measures.
Q3 **Explain** what the difference is between a 'qualitative risk assessment', a 'semi-quantitative' risk assessment and a 'quantitative' risk assessment.
Q4 **Explain** what a 'risk rating matrix' is.
Q5 **Explain** what the purpose of a Hazard Identification Study (HAZID) is.
Q6 **Explain** what the purpose of a Hazard and Operability Study (HAZOP) is.
Q7 **Identify** the kind of persons you would expect to be part of a HAZOP team.

Q8 **Explain** what a Failure Modes and Effects and Critical Analysis (FMECA) study is.

Q9 **Explain** what is meant by 'As Low As Reasonably Practicable' (ALARP).

Q10 The management of major incident risks should take a hierarchical approach. **Give** that hierarchical approach.

Q11 **Outline** the main principles for achieving an inherent safe design.

Q12 **Describe** the characteristics of:

(a) a pool fire
(b) a jet fire
(c) a flash fire or fire ball

Sub-element 1.4:
An organization's documented evidence to provide a convincing and valid argument that a system is adequately safe

Learning outcome

Explain the purpose and content of an organization's documented evidence to provide a convincing and valid argument that a system is adequately safe in the oil and gas industries.

🔑 **Key revision points**

Where documented evidence is used

The purpose of documented evidence such as safety cases and safety reports

The typical content of documents such as safety cases and safety reports

Where documented evidence is used

Safety case and safety reports

Organizations are obliged to produce documented evidence that their systems of operation at any of their installations or facilities are adequately safe. This document is known as a safety case or safety report. The requirement is to provide evidence and information that present a clear, comprehensive and defensible argument that a system is adequately safe to operate in a particular context. The document produced to provide this evidence for **offshore installations** is called a 'safety case'. For **onshore installations** the document is called a 'safety report'.

The purpose of documented evidence such as safety cases and safety reports

Purpose of documented evidence

Organizations are required to submit a safety case or safety report for each installation or facility they own or operate. This takes the form of a report which demonstrates the level of safety applied to that installation or facility. The safety case or safety report covers all aspects of health and safety on an installation/facility. It is submitted at the planning stage and remains in place throughout the lifespan of the installation/facility until it is decommissioned.

The typical content of documents such as safety cases and safety reports

Offshore safety case

The three overriding principles to be demonstrated in a safety case are:

1 The management system is adequate to ensure compliance with statutory health and safety requirements.
2 Adequate arrangements have been made for audit and for audit reporting.
3 All hazards with the potential to cause a major accident have been identified, their risks evaluated, and measures have been, or will be, taken to control those risks.

The underlying principles are as follows:

Factual information – This will include factual information about the installation itself, the plant and systems used, its location and external environment

Management of health and safety – This will show how the management system will apply appropriate levels of control during each phase of the installation's life cycle including the design, construction, commissioning, operation, decommissioning and dismantlement stages.

Control of major accident hazards – This will demonstrate that all hazards with the potential to cause a major accident have been, or will be, identified, their risks evaluated and that measures have been, or will be, taken to control those risks.

Major accident hazard identification – This will show how a systematic process has been, or will be, used to identify all reasonably foreseeable major accident hazards that are applicable to the installation.

Major accident risk evaluation – This will clearly show what criteria have been, or will be, adopted for major accident risk assessment.

Major accident risk management – This will describe what measures will be taken to manage major accident hazards.

Rescue and recovery – This will demonstrate that effective rescue and recovery arrangements have been planned for to cope with major accidents.

Life cycle requirements – This will describe how the principles of risk evaluation and risk management are being applied to the design to ensure that major accident risks will be controlled.

Combined operations – This will demonstrate how the management system addresses the additional risks associated with combined operations.

Decommissioning and dismantlement – When the installation is reaching the end of its working life, the safety case will have to be revised to deal with decommissioning or dismantlement operations.

Availability – The document should be made available to anyone on the installation.

Safe design concept – The report must include an explanation of how inherently safe design concepts were considered and applied.

Safety Management System (SMS) – This should demonstrate an appropriate level of control during each phase of the installation's life cycle. This will include the design, construction, commissioning and operation as well as the decommissioning and dismantlement phases.

Onshore safety report

The onshore safety report is split into five main sections:

Section 1: descriptive information

▷ An overview of the facility and its activities
▷ Information about dangerous substances in use at the facility
▷ Information about the surrounding environment

42

Section 2: information on management measures to prevent major incidents

▷ Major Accident Prevention Policy (MAPP)

Section 3: information on potential major incidents

Section 4: information on measures to prevent or mitigate the consequences of a major incident

Section 5: information on the emergency response measures of a major incident

▷ Onsite emergency plan
▷ Offsite emergency plan

Sub-element 1.4: An organization's documented evidence to provide a convincing and valid argument that a system is adequately safe

Learning outcome

Explain the purpose and content of an organization's documented evidence to provide a convincing and valid argument that a system is adequately safe in the oil and gas industries.

Revision exercise

Write your answers to the questions below on a separate sheet of paper without referring to the information in this book in the first instance. Once you have answered all of the questions, you can refer back to the revision guide to compare your answers and this will give you an indication of how much knowledge you have been able to absorb or whether you need to revise this section further.

Q1 **Explain** what the purpose of a safety case/safety report is.
Q2 An offshore safety case has **THREE** overriding principles which must be demonstrated within the document. **Outline** what those three principles are.

Q3 **Outline FIVE** of the underlying principles or issues which will be covered in an offshore safety case.

Q4 An onshore safety report is made up of **FIVE** main sections. **Describe** what each of those sections is and what issues they cover.

Element 2

CHAPTERS 5–10

Sub-element 2.1:
Contract management

Learning outcome

Explain the principles of assessing and managing contractors, including the roles of parties involved.

🔑 Key revision points

Scale of contractor use	☐
Contract management, ownership and site supervision/representation	☐
Contractor responsibilities	☐
Safe handover, understanding the hazards	☐

Scale of contractor use

Contractors are used extensively in the oil and gas industry for all kinds of work and at every stage of the operation. This extends to, and includes, exploration, construction, production and decommissioning, as well as the less technical services such as cleaning, catering and waste removal. They can also be used for emergency response.

Contractor management, ownership and site supervision/representation

Client (employing company)

These are those organizations which select and engage contractors to undertake specific work for them. They are obliged to protect their own employees, as well as the employees of the contractors they are engaging, from risks to health and wellbeing. The client also has a responsibility towards the contractor and the contractor's own staff for hazards that may occur as a result of the client's own activities. **The client is responsible for the workplace the contractor is working in**.

Contractors

These are those people or organizations who have been selected and engaged to undertake work for the client. **The contractor is responsible for the safe methods of working they use**.

Procedures for selecting suitable contractors

Clients need to ensure that any contractor they engage is competent. Determining a contractor's competence should be the first step in choosing a suitable contractor. The kind of issues which should be considered include the following:

▷ Is the contractor adequately insured?
▷ Has, or will, the contractor undertake a risk assessment of the proposed contracted work?

▷ Are the health and safety policies and practices of the contractor adequate?

▷ Is the contractor's recent health and safety performance reasonable (number of accidents, etc.)?

▷ Is the contractor's health and safety training and supervision adequate?

▷ Does the contractor have arrangements in place for consulting with their workforce?

▷ Does the contractor or their individual employees hold a 'passport' or other type of certification in health and safety training?

▷ What, if any, enforcement notices have been served on the contractor?

▷ Can the contractor offer any independent assessment of their competence?

▷ What references from previous employing companies can the contractor show?

▷ What relevant qualifications and skills does the contractor have?

▷ What is the level of competency of the staff doing the job?

▷ Is there appropriate certification for any equipment that the contractor might intend to use?

▷ What selection procedure does the contractor have for sub-contractors they might engage with?

▷ Is the contractor a member of a relevant trade or professional body?

▷ What is the contractor's financial viability?

▷ Does the contractor's safety method statement meet expectations?

Managing/supervising contractors

Once a contractor is selected and engaged, the client should provide an appropriate level of supervision of the work being undertaken by the contractor. Clients may need to agree with the contractor how the work will be done and the precautions that will be taken. This might include the following:

49

▶ What equipment should or should not be worked on/used
▶ Personal protective equipment to be used and who will provide it
▶ Working procedures, including any permit-to-work systems
▶ Training and induction in such systems
▶ The number of people needed to do the job
▶ The reporting of accidents and safe keeping of records and plans

Contractor responsibilities
Main responsibilities

It is the contractor who is responsible for the safe methods of working they employ. This is because, in many cases, they are being engaged because of their specialist knowledge and expertise, whereas the company contracting out the work may not have that expertise. Contractors also need to give an assurance that the work they carry out will not impact on the safety of operations (e.g. will not disable emergency systems such as alarms, emergency shutdown systems, etc.).

Other responsibilities of the contractor

The contractor also has other responsibilities. These include:

▶ **Risk assessment** – There is a requirement for a risk assessment to be conducted concerning the work which is to be undertaken by the contractor.
▶ **Instruction and training** – There is a requirement for contractors and subcontractors (as well as the client) to
   ▷ consider what information should be passed between them and agree appropriate ways to ensure this is done
   ▷ exchange clear information about any risks arising from their work including relevant safety rules and procedures as well as arrangements for dealing with emergencies
   ▷ provide any information, instruction and training to their employees on anything which may impact on their health and safety
   ▷ consult their employees (and safety representatives if applicable) on health and safety matters

50

Safe handover, understanding the hazards

Safe handover procedure

Whenever a site, or part of a site, piece of machinery or plant, is to be handed over to a contractor for them to undertake their work, a comprehensive system of safety checks should be implemented. This may include:

▷ Electrical isolations with lock offs as appropriate
▷ Mechanical isolations with lock offs as appropriate
▷ Physical barriers to restrict access to non-authorized personnel
▷ Pre-cleaning of the area
▷ Pre-cleaning of the equipment

There should be a record kept of what was included in the handover; particularly important are any isolations and barriers which may be involved.

Safe handback procedure

Once the contract is completed, there should be a procedure for handing the site, or part of a site, piece of machinery or plant, back to the client. This handback should follow an agreed procedure to ensure that all matters have been dealt with safely and securely. This will include confirming that all parts of the installation which have been temporarily the responsibility of the contractor are in full working order and also that any isolations and barriers have been removed.

Sub-element 2.1: Contract management

Learning outcome

Explain the principles of assessing and managing contractors, including the roles of parties involved.

✎ **Revision exercise**

Write your answers to the questions below on a separate sheet of paper without referring to the information in this book in the first instance. Once you have answered all of the questions, you can refer back to the revision guide to compare your answers and this will give you an indication of how much knowledge you have been able to absorb or whether you need to revise this section further.

Q1 In relation to contract management, **outline** what the main responsibility of the client is.

Q2 In relation to contract management, **outline** what the main responsibility of the contractor is.

Q3 When a client is going through the process of selecting a contractor there are a number of issues which can be considered to ensure the contractor is suitable. **Give SIX** of those issues.

Q4 When a site or piece of machinery is handed over to a contractor for them to undertake their work, a comprehensive system of safety checks should be implemented. **Outline** what these might include.

Sub-element 2.2:
Process safety management

Learning outcome

Outline the tools, standards, measurements, competency
requirements and controls applicable to Process Safety
Management (PSM) in the oil and gas industry.

Key revision points

The controls available in process safety management

Management of change controls

The controls available in process safety management

Process Safety Management (PSM)

PSM is the management of hazards associated with the processing of products, particularly highly volatile substances such as hydrocarbons. The principle behind it is to reduce the number of incidents involving the release of highly volatile or toxic substances or, at the very least, mitigate the severity of such incidents.

The key provision of process safety management is process hazard analysis. This is basically a careful review of what could go wrong and what safeguards must be implemented to prevent releases of hazardous substances.

Process safety management – process hazard analysis

The following issues must be addressed:

▶ The hazards of the process
▶ The identification of any previous incidents
▶ Engineering and administrative controls applicable to the hazards
▶ Consequences of failure of engineering and administrative controls
▶ The siting of the facility
▶ Human factors
▶ A qualitative evaluation of a range of the possible safety and health effects on employees in the workplace if there is a failure of controls

Process safety management – operating procedures

These should be consistent with the process safety information and provide clear instructions for safely conducting activities involved in each process. The procedures should address the following elements:

▶ Initial start-up
▶ Start-up following a turnaround, or after an emergency shutdown

▶ Normal operations
▶ Temporary operations
▶ Emergency operations
▶ Normal shutdown
▶ Emergency shutdown

Operating limits

▶ Consequences of deviation
▶ Steps required to correct, or avoid, deviation

Health and safety considerations

▶ Properties of, and hazards presented by, the products/chemicals used in the process
▶ Precautions necessary to prevent exposure, including engineering controls, administrative controls and personal protective equipment
▶ Control measures to be taken if physical contact or airborne exposure occurs
▶ Quality control for raw materials and control of hazardous chemical inventory levels
▶ Any special or unique hazards
▶ Safety systems (e.g. interlocks, detection or suppression systems) and their functions

Process safety management – employee participation

Management should consult with their employees and/or their representatives on the conduct and development of process hazard analysis and on the development of the other elements of process management.

Process safety management – training

For process safety management to be effective, it is necessary for all personnel involved in the process to be fully trained on the implementation of the safe working procedures which have evolved from the process hazard analysis procedure.

55

Process safety management – contractors

Clients need to ensure that any contractor they engage is competent.

Process safety management – pre-start-up safety review

Prior to the introduction of a hazardous product to a process, the pre-start-up safety review must confirm the following:

▶ Construction and equipment are in accordance with design specifications.
▶ Safety, operating, maintenance and emergency procedures are in place and are adequate.
▶ A process hazard analysis has been performed for new facilities and recommendations have been resolved or implemented before start-up, and modified facilities meet the management of change requirements.
▶ Training of each employee involved in operating a process has been completed.

Process safety management – mechanical integrity

It is important to maintain the mechanical integrity of critical process equipment to ensure it is designed and installed correctly and operates properly.

Process safety management – permits-to-work

A permit-to-work must be issued prior to any work or operations being conducted on or near a process. This includes any hot work, breaking of containment, confined space entry, working at height or over water, etc.

Process safety management – management of change

Any change to a process must be thoroughly evaluated in order to fully assess its impact on employee health and safety, as well as to determine what adjustments will be needed to operating procedures.

Process safety management – incident investigation

Each and every incident that did, or could have, resulted in the release of any flammable or toxic substance should be investigated.

Process safety management – emergency planning and response

If, despite the best planning, an incident does occur it is essential that an emergency plan has been pre-planned and is ready to be implemented.

Plant layout

With regard to safety, the most important factors of plant layout are those which:

▶ Prevent or mitigate an escalation of events
▶ Ensure safety of personnel within on-site buildings
▶ Control access of unauthorized personnel
▶ Facilitate access of emergency services

Plant layout – inherent safety

Possible methods to achieve an inherently safe design include:

▶ Reducing inventories wherever possible
▶ Replacing hazardous substances with less hazardous alternatives where possible
▶ Minimizing hazardous process conditions where possible, i.e. temperature, pressure, rate of flow, etc.
▶ Designing systems and processes to be as simple as possible
▶ Using failsafe design features

Plant layout – the Dow/Mond indices

These indices are useful in the development stage of a project because they evaluate process plant hazards and rank them against existing processes or projects in order to provide a comparative measure of the risk of fire and explosion. They do this by assigning them incident classifications and allowing objective spacing distances to be considered during the development phase of a process or project.

Plant layout – domino effects

When undertaking a hazard assessment of site layout it is essential that the consequences of loss of containment are fully evaluated. This includes the potential escalation of an incident and what this entails. These escalations are known as a domino effect.

Plant layout – fire

There are four ways a fire can spread:

- Direct burning
- Convection
- Radiation
- Conduction

Fire can be prevented from spreading by fire resistant walls, floors and ceilings.

However, running fires caused by flammable liquids can spread through drains, ducts and ventilation systems. Consideration of these possibilities should be given at the plant layout design stage. Furthermore, flammable gases and vapours may also find their way into passageways and cause a delayed ignition.

Plant layout – explosion

With regards to explosion pressure waves when designing plant layout, the following are mitigating factors:

- Ensuring separation distances are sufficient in even the worst case so that damage to adjacent plants will not occur
- Providing blast walls or locating adjacent plant in strong buildings
- Ensuring the walls of vessels are thick and strong enough to withstand a pressure wave from an explosion
- Ensuring any explosion relief vents are directed away from vulnerable areas, e.g. other plant or buildings, or roadways near site boundaries

Plant layout – toxic gas releases

The release of toxic gas may render a plant or process inoperable due to the domino effect which may ensue. This may be prevented or mitigated by:

▷ The implementation of automatic control systems (which use inherently safer principles)
▷ Controlling plant and processes from a remote facility, e.g. a suitably protected control centre

Plant layout – reduction of consequences of event on and off site

With regard to the design of inherently safe plant layout, other measures which may be considered include the following:

▷ Storage of flammable/toxic material away from process areas.
▷ Siting hazardous plant and processes away from main roadways within the site.
▷ Fitting remote-actuated isolation valves where large inventories of hazardous materials may be released.
▷ Using the terrain as a means of controlling potential releases of liquid hazardous material. This can include embankments (bunds), ditches, dykes, etc.
▷ The siting of plant within buildings to act as a secondary containment facility.
▷ Where there is the potential for minor release of flammable gases or vapours, then siting this plant in the open air. This will ensure there is a rapid dispersion of these gases or vapours.
▷ Classification of areas where flammable gases, vapours and dusts may be released. This will enable ignition sources to be controlled and eliminated.

Plant layout – positioning of occupied buildings

The distance between, and the position of, occupied buildings and buildings housing plant will be governed by the need to reduce the domino effect of a fire, an explosion or the release of toxic gases. Over and above the distance and proximity issues, consideration

59

should also be given to evacuation routes which should not be impeded by poor plant layout. The siting of occupied buildings should generally be upwind of hazardous plant areas.

Plant layout – aggregation/trapping of flammable vapours

In order that flammable/toxic vapours are not allowed to accumulate in buildings and create a hazardous event, buildings should be well ventilated by natural or engineered ventilation.

Storage of flammable products should be in the open air so that any minor leaks can be allowed to disperse naturally, although some form of warning system should be deployed so that a minor leak does not develop into a major leak.

Control room design

With regards to control room design, there are two main aspects that should be part of the design consideration. These are:

▷ The ability of the control room to withstand a major hazardous event such as a fire, explosion or release of toxic gas or smoke
▷ The efficient and appropriate layout of the control room and its equipment to ensure the effective operation and control of the plant under any circumstances, including an emergency

Control room design – control room structure

The structure of the control room should be such that it can be safely used by personnel to maintain plant control at all times, even if an emergency situation or undesirable event takes place. Events that have the potential to affect the control room are:

▷ Vapour Cloud Explosion (VCE)
▷ Boiling Liquid Expanding Vapour Explosion (BLEVE)
▷ Pressure burst
▷ Toxic gas release
▷ Fires, including pool fires, jet fires, flash fires and fire balls

Temporary refuge integrity can be described as the ability to protect the occupants following a hazardous event for a specific time period such that they will remain safe until they decide there is either a need to evacuate the installation, or to recover the situation.

The design of a Temporary Refuge (TR) should take into account potential hazards and allow personnel access to a safe evacuation route. This means that, should a facility be evacuated, the route to safety should take personnel away from areas where plant and hazardous materials are stored, as well as away from any potential exposure to toxic gases or fumes.

Management of change controls

Management of change

Many accidents and incidents, some of them catastrophic, can be attributed to changes in processes and equipment. Management should have systems in place to ensure that any proposed changes are evaluated before they are implemented. The management of change procedure should:

▷ Include expert personnel to review the proposed changes to ensure that they will not result in any operations exceeding established operating limits
▷ Ensure that any proposed changes are subject to a safety review using hazard analysis techniques
▷ Have in place arrangements for the control of relevant documents
▷ Ensure that any changes in the operating envelope (e.g. temperatures, pressures, flow rates, etc.) are communicated to the operators and documented

Sub-element 2.2: Process safety management

Learning outcome

Outline the tools, standards, measurements, competency requirements and controls applicable to Process Safety Management (PSM) in the oil and gas industry.

✎ Revision exercise

Write your answers to the questions below on a separate sheet of paper without referring to the information in this book in the first instance. Once you have answered all of the questions, you can refer back to the revision guide to compare your answers and this will give you an indication of how much knowledge you have been able to absorb or whether you need to revise this section further.

Q1 **Explain** what process safety management is and what its main aim is.

Q2 The key provision of process safety management is process hazard analysis. **Explain** what this is and what its main aim is.

Q3 When considering plant layout, **outline** the most important safety factors to be considered.

Q4 **Outline** what sort of information the Dow/Mond indices offer which might assist in plant layout design.

Q5 **Explain** what is meant by the 'domino effect' when considering plant layout design.

Q6 With regard to control room design, **outline** the two main aspects that should be part of the design consideration.

Q7 **Explain** what is meant by the phrase 'temporary refuge integrity'.

Q8 **Outline** what aspects management of change controls should include to ensure changes to process and equipment are implemented as safely as possible.

Sub-element 2.3:
Role and purpose of a permit-to-work system

Key revision points

Role and purpose of a permit-to-work system ☐

Key features of a permit-to-work system ☐

Types of permit ☐

Interfaces with adjacent plant ☐

Interfaces with contractors ☐

Lock out, tag out and isolation ☐

Role and purpose of a permit-to-work system

A permit-to-work is a detailed document which describes specific work at a specific site at a particular time which is to be carried out by authorized personnel. It also sets out any precautions and control measures which are necessary to complete the work safely.

The key features of a permit-to-work system

The essential features of permit-to-work systems are:

▶ The clear identification of who is responsible for specifying any necessary precautions
▶ The identification of personnel who may authorize particular jobs (including any limitations to their authority)
▶ The clear identification of any work classified as hazardous
▶ Clear identification of:
  ▷ tasks
  ▷ risk assessment
  ▷ duration of permitted tasks
  ▷ additional or simultaneous activity
  ▷ control measures
▶ Training and instruction regarding the issue, use and closure of permits-to-work

▷ Monitoring and auditing to ensure that the system is working as planned

A permit-to-work document also provides written evidence, by means of a signature, that at various stages during the time the permit-to-work was live, each stage was accepted by the person holding the permit-to-work.

The **objectives and functions of a permit-to-work system** are as follows:

▷ It ensures that proper authorization of designated work has been granted.
▷ It ensures those people who are conducting the work know the exact nature of the task including hazards, restrictions, time limitations, etc.
▷ It specifies the controls and precautions necessary to undertake the work safely, e.g. isolating machinery.
▷ It ensures those in charge of the location are aware the work is being carried out.
▷ It provides both a system of continuous control and a record that appropriate precautions have been considered and applied by competent persons.
▷ It affords the ability to display, to those who need to know, exactly what work is ongoing.
▷ It provides a procedural means of suspending work when this is necessary.
▷ It provides an ability to control work which might interact or conflict with ongoing operations or other permit-to-work activities.
▷ It provides a procedural means of handing over the work when that work covers more than one shift.
▷ It provides a procedural means of handing back the area or plant which has been involved in the work.

Appropriate use of permit-to-work systems

Permits-to-work should be applied whenever work is to be carried out which may affect the health and safety of personnel, plant or the environment. This does not normally include work which is routine

65

(e.g. routine maintenance which is carried out in non-hazardous areas). They should be considered, however, for:

▶ Special operations (i.e. work that is not routine)
▶ Work which is done other than normal production work (e.g. inspection, testing, dismantling, modification or adaptation of processes as well as repair work and non-routine maintenance)
▶ Work done by two or more individuals or teams where activities need to be co-ordinated to ensure the work is completed safely
▶ Work which will take longer to complete than one shift and the work and responsibility for it needs to be formally handed over

Display

Copies of a permit-to-work need to be clearly displayed:

▶ At the place where the work is to be carried out or in a recognized location close by. If the work is to be carried out in several locations, the permit should be kept by the performing authority.
▶ In the main control room or permit co-ordination room as well as in local control rooms.

Suspension

Sometimes work has to be suspended, for example:

▶ If there is a change to the planned type or extent of work
▶ If there is a risk that the work to be carried out will present a hazard. For example when the permit-to-work is for hot work and at the same time there is a need to carry out sampling of process fluid or gas which itself poses the risk of the release of a dangerous substance
▶ Whilst awaiting the delivery of spares
▶ Where the work is in conflict with other work being carried out

The permits which have been suspended should still be kept on the permit recording system. They should specify the condition of the plant where the work has been suspended and any consequences for other activities which may be affected by that condition.

Permit interaction

When a permit-to-work is being considered for a job, it is important that the person issuing the permit be aware of any other activity (either planned or already underway) which may interact with the work to be done under the permit.

Handover

A handover procedure is required in the event of any work being done under a permit-to-work which carries over into another shift. This allows the outgoing shift to communicate all relevant information about the work and the conditions of the permit to the new shift to ensure continuity and safety. The information will include:

▷ Any outstanding work under permit control
▷ The status of that work (this should be left in a condition which can be reliably explained to and understood by the incoming shift)
▷ The status of any other work which may affect the permit controlled job (e.g. isolations)
▷ The status of the plant or installation

The information can be recorded in a permit log, permit file or on display boards.

Handback

The handback procedure should be a process of reinstating a plant or installation (or the part that has been worked on under a permit). This will ensure that the work has been completed and is in a safe condition.

Permit authorization and supervision

A permit-to-work must be co-ordinated or controlled by the issuing authority (or other responsible authority). This includes the monitoring and supervision of the work to ensure that it is being done according to the specific procedures detailed on the permit.

67

Permit-to-work template

Permit-to-Work	
1) Permit Title	2) Permit Number
	Reference to other relevant permits or isolation certificates
3) Job location	
4) Plant identification	
5) Description of work to be done and its limitations	
6) Hazard identification Including residual hazard and hazards associated with the work	
7) Precautions neccessary People who carried out precautions, e.g. isolating authority, should sign that precautions have been taken	
8) Protective equipment	
9) Authorization Signature (issuing authority) confirming that isolations have been made and precautions taken, except where these can only be taken during the work. Date and time duration of permit. In the case of high hazard work, a further signature from the permit authorizer will be needed	
10) Acceptance Signature confirming understanding of work to be done, hazards involved and precautions required. Also confirming permit information has beeen explained to all permit users	
11) Extension/shift handover procedures Signatures confirming checks made that plant remains safe to be worked upon, and new performing authorities and permit users made fully aware of hazards/precautions. New expiry time given	
12) Handback Signed by performing authority certifying work completed. Signed by issuing authority certifying work completed and pant ready for testing and re-commisioning	
13) Cancellation Certifying work tested and plant satisfactorily re-commisioned	

Figure 7.1   Permit-to-work template

Source: Wise Global Training interpretation based on HSE publication HSG250.

Types of permit

Permits-to-work cover many different types of operations and tasks, and the following are examples of types of job where permits should be considered:

▶ Work where heat is used or is generated, for example welding, grinding, etc.

▶ Work which involves breaking containment of a flammable or dangerous substance
▶ Work which involves breaking containment of a pressure system
▶ Work on electrical equipment
▶ Work within tanks and other confined spaces
▶ Working at height
▶ Work involving hazardous substances
▶ Well intervention
▶ Diving operations
▶ Work involving pressure testing

Interfaces with adjacent plant

Any permits-to-work being considered must be cross-referenced to any other work either being planned or in progress. This is to ensure there is no conflict or interaction which may result in the creation of additional hazards for any of the work.

Interfaces with contractors

A comprehensive induction procedure should be undertaken with regard to permit-to-work systems with all contractors prior to any work being undertaken. This should include:

▶ Ensuring the contractor and his/her employees fully understand the permit-to-work systems and the arrangements within them which apply to the site where they are to carry out the work
▶ Ensuring that all personnel from the performing authority and other users are fully trained and aware of any specific arrangements in force to make the job safe at the location where they are to carry out the work
▶ Ensuring the contractor understands fully the principles of the permit-to-work systems within the industry

69

Lock out, tag out and isolation

The term 'lock out' or 'tag out' refers to a procedure which is aimed at safeguarding someone who is working on or near plant or machinery, from that plant or machinery unexpectedly starting up or releasing energy of some kind.

This procedure involves an authorized person to disconnect the plant or machinery from its energy source. This person then locks or tags the isolator in order to prevent anyone from re-energizing the plant or machine.

Basic procedure for implementing lock out and tag out actions

1   The person authorized to conduct the lock out or tag out action will identify the source(s) of energy to be controlled and the method of control to be used.
2   All personnel who may be affected by the isolation caused by the lock out, tag out action should be informed.
3   The process or system should be shut down as normal and confirmation made that all controls are in the off position and that all moving parts have stopped.
4   The process or system should be isolated from its source of energy.
5   Verification that the isolation is effective needs to be made. This is normally done by making an attempt to try to restart the system or process. If it does not start, the isolation has been effective.
6   Once isolation is verified the start-up controls must be returned to their off or neutral position.
7   Once the work is completed, the lock out, tag out device will be removed.

Sub-element 2.3: Role and purpose of a permit-to-work system

Learning outcome

Explain the role and purpose of a permit-to-work system.

Revision exercise

Write your answers to the questions below on a separate sheet of paper without referring to the information in this book in the first instance. Once you have answered all of the questions, you can refer back to the revision guide to compare your answers and this will give you an indication of how much knowledge you have been able to absorb or whether you need to revise this section further.

Q1 **Describe** the role and purpose of a permit-to-work system.

Q2 **Outline** the key features of a permit-to-work system.

Q3 **Outline** the objectives of a permit-to-work system.

Q4 **Describe** the type of work you would expect a permit-to-work system to cover.

Q5 With regard to other work that may be ongoing at the same time as a permit-to-work is being issued, **explain** what considerations should be made.

Q6 With regard to the issuing of permits-to-work to contractors, **explain** what considerations should be made.

Q7 **Describe** what the term 'lock out' or 'tag out' refers to and what its aim is.

CHAPTER 8

Element 2

Sub-element 2.4:
Key principles of safe shift handover

Key revision points

Placing greater reliance on written communication between handover of 12-hour shifts

Two-way communication at handover with both participants taking joint responsibility

What shift handover should include

Placing greater reliance on written communication between handover of 12-hour shifts

Shift handover

Where a production process or operation continues indefinitely there inevitably comes a time when those personnel undertaking or overseeing the process or operation have to leave their duties and let others take their place. This is when 'shift handover' takes place. This can be defined as **the effective transfer of information between the outgoing and incoming parties with no miscommunication or misunderstanding**.

Shift handover – background

Oil and gas exploration and production processes are continuous, 24-hour operations. The personnel who work on these operations and processes tend to work alternate 12-hour shifts over a 2- to 4-week period. These operations tend to be highly complex and place demands on the operator's skills as far as information processing and decision-making is concerned. Also, processes may evolve over many hours, or even days, and the ability of the operator to mentally understand what is going on, or what is expected, is critical. Successful control of an operation requires three elements:

▷ A clear understanding of the expected outcomes of the production process or operation
▷ A clear understanding of the current state of the production process or operation
▷ A clear understanding of the dynamics of the production process or operation

Two-way communication at handover with both participants taking joint responsibility

During shift handover, the communication of information from one person to another is a two-way process. The person who is providing the information needs to be sure that the person who is receiving

74

it is in no doubt about the message being conveyed. It is a mutual interaction between the two parties – person(s) A (the sender of the information) and person(s) B (the person(s) receiving the information). Person(s) A needs to receive feedback from person(s) B confirming that the information has been received and correctly interpreted in order for the two-way communication to have been successful.

What shift handover should include

Why good shift handover is important

Good shift handover allows the accurate and effective communication of vital information to incoming personnel in order to enable the safe operation of process plant and equipment. The aim of shift handover is to communicate effectively all relevant information to allow the continuation of safe and efficient systems and process operations. There are three elements which contribute to an effective shift handover:

1. A period of time when the outgoing team prepares the information it will be conveying to the incoming team.
2. A period of time when both the outgoing and incoming team communicate with each other and exchange all relevant information.
3. A period of time when the incoming team cross-checks the information passed onto it as it takes on the responsibility for ongoing operations.

Key principles in effective shift handover

Management should ensure that the facilities, the arrangements and the infrastructure allow for good shift handover practice. This should:

▷ Provide all staff who are involved in the handover process with training and the development of good communication skills
▷ Provide written procedures for effective shift handover
▷ Emphasize the importance of effective shift handover
▷ Ensure that both parties take joint responsibility for conducting an effective shift handover

75

▶ Emphasize the use of both written and verbal means of communication

▶ Ensure that any important information is written into a log or report form. The log or report form should be brought to the attention of personnel at handover

▶ Include all personnel from both the incoming and outgoing team in the handover without exception

▶ Ensure that managers and supervisors are available at all times and implement an 'open door' policy

Shift handovers should involve the following principles:

▶ Be treated as high priority

▶ Not be rushed but be allowed as much time and resources as are necessary to ensure the accurate communication of information

▶ Be conducted using both verbal and written means of communication

▶ Be conducted face to face with both parties taking joint responsibility for the effective communication of necessary information

▶ Be conducted in an environment which is conducive to good communication without distractions

▶ Involve all shift personnel

As far as the operation of the plant and equipment is concerned, issues covered should include:

▶ Work permits – the status of existing permits and the status of work in progress

▶ The updating of work permits

▶ Preparations for upcoming maintenance

▶ New personnel to the shift

▶ Any plant overrides – existing and planned

▶ Information about any abnormal events

▶ Any existing or planned shutdowns

▶ Any changes in plant parameters

▷ Any routine operations and existing parameters which may need to be carried out by personnel from the incoming shift

▷ Any breakdowns which may have occurred

▷ Any faults which have occurred with safety critical equipment

▷ Inhibits to the Fire and Gas (F&G) and Emergency Shutdown (ESD) systems

▷ Any completed work and equipment which has returned to service

Sub-element 2.4: Key principles of safe shift handover

Learning outcome

Explain the key principles of safe shift handover.

✎ Revision exercise

Write your answers to the questions below on a separate sheet of paper without referring to the information in this book in the first instance. Once you have answered all of the questions, you can refer back to the revision guide to compare your answers and this will give you an indication of how much knowledge you have been able to absorb or whether you need to revise this section further.

Q1 **Outline** what is involved in a shift handover process.

Q2 The communication of information during shift handover is a two-way process. **Describe** this two-way process.

Q3 Effective shift handover involves three specific periods of activity. **Describe** what these periods of activity involve.

Q4 Shift handover involves a number of important principles. **Outline** what these principles are.

Q5 **Outline** what kind of issues might be covered in shift handover involving plant and equipment.

CHAPTER 9

Element 2

Sub-element 2.5: Plant operations and maintenance

Learning outcome

Explain the importance of safe plant operation and maintenance of hydrocarbon-containing equipment and processes.

🔑 Key revision points

Asset integrity, including inspection, testing, maintenance, corrosion prevention, competency and training ☐

Risk-based maintenance and inspection strategy ☐

Techniques, principles and importance of safe operation, standard operation procedures and maintenance ☐

Control of ignition sources during maintenance and operations ☐

Cleaning and gas freeing; purging; venting; draining of water, product, oxygen and non-condensables and inerting ☐

Asset integrity, including inspection, testing, maintenance, corrosion prevention, competency and training

Asset integrity

In the oil and gas industry, offshore installations, refineries, storage terminals and pipelines are often referred to as 'assets'. Maintaining an asset's ability to function effectively and efficiently without creating undue hazards to any persons or the environment is essential and this is referred to as maintaining an 'asset's integrity'.

Managing an asset's integrity involves ensuring the people, the systems, the processes and the resources are all available and used when required in order to maintain integrity.

The management of an asset's integrity will cover its entire life from initial design through to commissioning and operating and, finally, to decommissioning.

Safety critical elements

Because of the raised risk level of working with highly flammable, hydrocarbon based substances, it is essential that safety devices are incorporated within an asset so that if a hazardous situation or incident does occur there are devices in place to deal with the consequences by quelling, controlling or mitigating the event or situation. These are known as 'safety critical elements'. They include:

Safety critical elements – blowout preventers

These are used to control blowouts. When a blowout – i.e. a sudden surge of pressure from within the drill hole – occurs, the blowout preventer is automatically activated by the ensuing pressure, making sure that the pressure is contained and the hole is sealed.

80

Safety critical elements – fire deluge systems

These are positioned where the risk of fire is a serious concern. They can apply a continuous and high volume of water to an area which includes the hazard as well as possible escape routes for personnel.

Safety critical elements – emergency shutdown valve

This is a device which is designed to automatically shut down the flow of a fluid when a dangerous situation is detected.

Safety critical elements – fire and gas detection systems

Fire detectors are used to detect heat, flames and smoke, whilst the gas detectors are used to detect flammable and toxic gas as well as vapours.

Safety critical elements – safety inspection and testing

It is essential that all safety critical elements are 100 per cent effective at all times, and the only way to be assured of this is to conduct a pre-defined programme of inspection, testing and maintenance.

Safety critical elements – corrosion prevention

When it comes protecting material from corrosion, the use of cathodic protection systems and the injection of chemical corrosion inhibitors into pipework and process systems are means of controlling corrosion. This should be done in conjunction with inspection, testing and cleaning on a regular basis.

Safety critical elements – training and competency

It is essential that personnel have the appropriate training in order to become competent in the skills necessary to maintain asset integrity.

Risk based maintenance and inspection strategy

Risk based maintenance

This is a technique based on increasing the reliability of equipment by assessing the probability of various failure scenarios and applying an appropriate maintenance schedule in order to pre-empt these failures.

Risk based maintenance – inspection

In order to ensure the integrity of plant and equipment during its service life, a regulated inspection strategy should be implemented to establish that:

▶ No damage has occurred to the plant or equipment
▶ Wear and tear are being kept under review and any components that are considered to be in need of replacement are notified to the maintenance crew

Condition monitoring maintenance

This involves taking readings of equipment in ways that allow those readings to give an insight into the condition of the plant or equipment without having to shut it down or dismantle it. For example, equipment which has rotating components can be monitored using vibration analysis.

Techniques, principles and importance of safe operation, standard operation procedures and maintenance

Pre-start-up safety reviews

A pre-start-up safety review is conducted:

▶ Before the actual start-up of a new installation
▶ When new chemicals or other hazardous materials are introduced into a process
▶ When existing facilities have had significant modifications or a maintenance shutdown

82

The review is conducted to ensure that:

▶ All materials and equipment which have been used to construct/modify/maintain the process system conform to the design criteria.

▶ All of the following have been inspected, tested and certified:
  ▷ process systems and hardware (including computer control logic)
  ▷ alarms and instruments
  ▷ relief and safety devices and signal systems
  ▷ fire protection and prevention systems.

▶ Safety, fire prevention and emergency response procedures have been formulated, reviewed and tested and are appropriate and adequate.

▶ Start-up procedures are in place and all appropriate actions have been taken.

▶ A process hazard analysis has been conducted and any resulting recommendations have been implemented or resolved and any actions taken have been documented.

▶ All training (initial, induction and/or refresher) has been undertaken for all personnel.

▶ All procedures have been completed and are in place, including: any management of change requirements for new processes; and any modifications to existing processes have been met.

Control of ignition sources during maintenance and operations

Sources of ignition include:

▶ Cigarettes and matches
▶ Heated process vessels, e.g. dryers and furnaces
▶ Flames from cutting and welding
▶ Lightning strikes
▶ Electric sparks from electric motors and switches
▶ Electromagnetic radiation of different wavelengths
▶ Electrostatic discharge sparks
▶ Impact sparks, friction heating or sparks

▶ Direct fired space or process heating
▶ Vehicles (unless they have been specially designed or have been adapted)

All areas designated as hazardous need to have systems and control measures in place to control ignition sources. These control measures should include:

▶ The implementation of a permit-to-work system for maintenance activities which may generate sources of ignition, e.g. grinding, use of a blowtorch, etc.
▶ There should be strict working procedures in place during periods when hazardous activities are being conducted, e.g. during the loading and unloading of tankers.
▶ Prohibition of smoking and the use of matches/lighters.
▶ Only using electrical and mechanical equipment and instrumentation which has been designed and manufactured for use in the zones in which it is to be used.
▶ Controlling the risk from pyrophoric scale (this is usually associated with the formation of ferrous sulphide within process equipment).
▶ The appropriate selection of vehicles and internal combustion engines for zoned areas in which they are to be used.
▶ The earthing of all plant and equipment.
▶ All equipment should be electrically bonded so there is no difference in electrical potential between equipment which might cause a static electrical spark.
▶ The installation of lightning protection, which includes having snuffing systems on vent stacks.
▶ The elimination of surfaces which are above the auto-ignition temperature of any flammable substances or materials which are being handled or stored.

Cleaning and gas freeing; purging; venting; draining of water, product, oxygen and non-condensables and inerting

Gas freeing and purging operations

This involves filling spaces in plant or equipment, currently filled with vapour, with an inert gas, sometimes known as a blanket gas. This enriches the atmosphere in the vapour space with inert gas in order that the atmosphere is taken to a level below its lower flammable limit.

Inerting operations

The partial or complete substitution of a flammable or explosive atmosphere in a contained environment by an inert gas is a very effective means of preventing an explosion. Inerting is typically used in storage tanks where a material may be above its flashpoint. Inert gases are also used in the transfer of flammable liquids under pressure, such as the transfer of hydrocarbons from seagoing oil tankers to land based facilities.

Venting operations

During any process operation, there will always be a need to vent the system if an overpressure problem occurs, which will then activate a pressure relief valve. The venting system is made up of a series of pipes which are connected to each pressure relief valve. These pipes then channel the overpressure through to the vent stack where the overpressure is released to atmosphere in a safe and controlled manner (blown down). Alternatively, the overpressure can go to a flare where it is allowed to combust (burn) in a controlled manner.

Drained water

All hydrocarbon processes produce a mixture of oils, gases, solids and water, all of which require to be separated so they can be dealt with individually. This separation process takes place in numerous pieces of equipment such as inlet separators, production separators, 3-phase separators, slug catchers, surge drums, filters, etc.

Sub-element 2.5: Plant operations and maintenance

Learning outcome

Explain the importance of safe plant operation and maintenance of hydrocarbon-containing equipment and processes.

 Revision exercise

Write your answers to the questions below on a separate sheet of paper without referring to the information in this book in the first instance. Once you have answered all of the questions, you can refer back to the revision guide to compare your answers and this will give you an indication of how much knowledge you have been able to absorb or whether you need to revise this section further.

Q1 **Explain** what is meant by the phrase 'asset integrity'.

Q2 **Describe** what is meant by the phrase 'safety critical elements'.

Q3 **Give FOUR** examples of safety critical elements.

Q4 **Describe** what 'risk based management' is.

Q5 **Describe** what 'condition monitoring maintenance' is.

Q6 **Give THREE** examples of situations when a pre-start-up review should be conducted.

Q7 **Explain** what gas freeing and purging operations involve.

Q8 **Explain** what inerting operations involve.

Q9 **Explain** what venting operations involve.

Sub-element 2.6:
Start-up and
shutdown

Learning outcome

Outline the hazards, risks and controls to ensure safe start-up and shutdown of hydrocarbon-containing equipment and processes.

Key revision points

Hazards and controls associated with safe start-up and shutdown

Hazards and controls associated with water and hydrates, their presence and removal

Hazards and controls associated with testing, commissioning and hook up

Hazards and controls associated with safe start-up and shutdown

Safe start-up and shutdown

The most hazardous part of operating any plant or process system is the starting up or shutting down procedure of that system. This is because the system is constantly evolving until it reaches its optimum level.

Safe start-up and shutdown – operating instructions and procedures

The 'operating procedure' document should take into consideration the following points:

▶ There should be no easier or more dangerous alternative procedure other than that specified in the operating procedure documentation.

▶ A review system should be in place to ensure that the operating procedures are kept up to date, and also to detect any errors in procedures which can be quickly corrected.

▶ The personnel involved in the operation should be involved at the design stage of the operating procedures to ensure that the document does not become too prescriptive.

▶ Operating procedures should include information about the requirements necessary for the use of personal protective equipment when carrying out the operating procedures.

▶ At the start of the procedure the document should include:
  ▷ an overview of the work to be done
  ▷ any risks which the operator may be exposed to, based on a risk assessment of the work
  ▷ any prerequisites, which should be clearly stated in order that a check can be made to ensure that it is safe to begin the procedures.

▶ Each operating procedure document should be dated and an expiry date indicated where appropriate, e.g. stating that it is valid for six months from the date given on the document.

▷ The document should make it clear which procedures apply to which situations. There should be no ambiguity – an appropriate method of coding each procedure should be used.
▷ The document should:
  ▷ ensure that the most important information on the page is clearly identified and made the most prominent on the page.
  ▷ ensure different sub-tasks are listed under separate headings to differentiate them.
  ▷ be written in a language which is simple and familiar to the operators who will be performing the tasks.
  ▷ ensure that the nomenclature (terminology) used is consistent with that used on controls or panels.
  ▷ ensure that any warnings, cautions or notes are put immediately before the instruction step to which they refer.
  ▷ ensure that any shapes, symbols and colours which are used for graphics are consistent and conform to industry standards.

Safe start-up and shutdown – thermal shock

Thermal shock is where a material is exposed to a sudden and significant change in temperature. This results in the material expanding at different rates within a limited area, causing a crack or failure.

Thermal shock can be reduced by:

▷ The gradual introducing of steam or warm product from a lower temperature base (i.e. not superheated steam at the outset)
▷ Thoroughly warming up the systems prior to use
▷ Designing in expansion loops into the system
▷ Using materials with greater thermal conductivity
▷ Reducing the coefficient of expansion of the materials
▷ Increasing the strength of the materials

Safe start-up and shutdown – shutdown

Shutting down a process can be equally as hazardous as starting it up. Consequently, the personnel involved should work as a team from a predetermined operating procedure which will take into account all possible eventualities associated with shutdown.

89

Hazards and controls associated with water and hydrates, their presence and removal

Water and hydrates – their presence and removal

When the flow of product or the pressure of the product is reduced, this has the effect of reducing the temperature, and it is at this point when the formation of a hydrate can occur if there is water within the system.

Hydrates can also be formed in pipelines which run across the sea bed. This is because temperatures at sea bed level can be low enough to freeze un-salted water which accompanies the oil and gas as it is extracted from the wells.

Some hydrates are made up of additional materials, as well as water, and these can form into sticky crystals which have the potential to grow into large ice plugs which can completely block a pipeline.

Water and hydrates – controls

To prevent hydrates forming in the first place, antifreeze (usually methanol or glycol) can be introduced into the process system so that it forms part of the product. This antifreeze is usually recovered later when the oil and gas arrives onshore, so it can be reused.

Removal of water from processed oil and gas can be achieved by a number of methods. Each has its own advantages and disadvantages.

Removal of water – gravity separation

When a product which is a mixture of oil and water is allowed to stand for a while, water will naturally settle to the bottom as the specific gravity of water is greater than that of oil. This can then be drained off.

Removal of water – centrifuge (spin the oil clean)

This system uses the principle of centrifugal force and the fact that water has a different specific gravity from oil, and that difference allows this system to separate the two materials.

Removal of water – absorption removal

Absorption removal is the process of using filters to absorb the moisture from product as the product passes through the filter.

Removal of water – vacuum dehydration

The process of vacuum dehydration works on the principle that water boils at a lower temperature when it is at a lower pressure. Consequently, at 0.9 bar, water will boil at around 52°C. Vacuum dehydration units reduce the pressure of the product within it. Air which has been dried and warmed is then passed over the product and the moisture is then transferred to the air from the product in the form of steam vapour.

Removal of water – air stripping

Air stripping is another form of vacuum dehydrator which works by mixing air, or nitrogen gas, with a stream of heated product within the air stripping unit. The gas then absorbs the moisture from the product and when the gas and product mixture is expanded, the gas separates from the product and takes the moisture with it.

Removal of water – heating the oil dry

Some processes, because they run at elevated temperatures, are self-cleansing due to the fact that water naturally evaporates at these temperatures.

Hazards and controls associated with testing, commissioning and hook up

Commissioning

Commissioning a process plant involves undertaking tests on the plant prior to it going into production in order to determine that it will function adequately and safely. The commissioning process also includes training both the people who will operate the plant and the control room staff who will oversee operations. Finally, it includes writing

91

the operating procedure document. A typical sequence of phases undertaken during the commissioning process is set out below.

1 The system configuration is checked (walking the line).
2 The pipework and system integrity is checked.
3 The instrumentation system is checked.
4 All alarms are verified as working.
5 All lines and vessels are flushed and cleaned.
6 All ancillary equipment is inspected and assessed as to its adequacy.
7 All instruments and vessels are calibrated.
8 The start-up protocol is established.
9 The shutdown protocol is established.
10 Commissioning trials are undertaken.
11 The plant is hooked up.
12 The plant is handed over.

Sub-element 2.6: Start-up and shutdown

Learning outcome

Outline the hazards, risks and controls to ensure safe start-up and shutdown of hydrocarbon-containing equipment and processes.

✎ Revision exercise

Write your answers to the questions below on a separate sheet of paper without referring to the information in this book in the first instance. Once you have answered all of the questions, you can refer back to the revision guide to compare your answers and this will give you an indication of how much knowledge you have been able to absorb or whether you need to revise this section further.

Q1 **Outline** what the 'start-up and shutdown operating instructions and procedures' document should take into consideration.
Q2 **Explain** what 'thermal shock' is and how its effects can be reduced.

Q3 **Explain** how hydrates can be controlled within a process system.

Q4 **Outline THREE** methods of removing water from processed oil and gas.

Q5 **Explain** what 'commissioning' a process plant involves.

Element 3

CHAPTERS 11–16

CHAPTER 11

Element 3

Sub-element 3.1:
Failure modes

Learning outcome

Outline types of failure mode that may lead to loss of containment
of hydrocarbons.

🔑 Key revision points

Failure modes introduction and terminology

Failure modes

What is meant by a safe operating envelope

Use of knowledge of failure modes in initial design, process and
safe operating procedure

Failure of the annular rim (bottom rim of storage tank)

Failure modes: introduction and terminology

There is a wide range of materials used in the construction of facilities in the oil and gas process industry. The properties required of materials to cope with the wide variation of conditions can be extensive and the process of considering which material is best for any given situation is a complex one. The first step in this approach is to understand the different ways in which the failure of materials can occur and what controls can be put in place to counteract these failures.

Tension is a force which is acting in two opposite directions. It could also be described as trying to stretch the material or pull the material apart.

Compression is the opposite of tension in that the material is being compressed by pressure.

Tension and compression are two forces which can be applied to one piece of material at the same time. For example, if a bending motion is exerted on a steel bar, this creates both tension and compression forces within the bar.

Shear is where two forces are being applied to a material but in opposite directions. The shear force acts as if it is trying to tear the material apart.

Properties of materials

Each material has its own properties, and understanding what these properties are will help us understand which materials are best applied to specific situations.

Properties of materials – ductile

A material which is ductile is one which can be subjected to tensile forces without it fracturing. An example of this is where a wire is drawn through an extrusion die.

Properties of materials – malleable

A material which is malleable is one which can have its shape changed without it cracking. This could be through compression (hammering) or by bending. Lead, tin and copper are all malleable materials.

Properties of materials – brittleness

A material which is brittle is one which has no plastic deformation characteristics. This means that if a force is applied which exceeds its characteristic threshold, it will crack or shatter with very little deformation. Cast iron is an example of a brittle material.

Properties of materials – elasticity

A material which has elasticity is one which can resume its former shape or dimension after a deforming force is applied and then released.

Failure modes

Creep

This is where a solid material is subject to long term exposure of high stress levels and gradually deforms in shape or dimension. Creep is exacerbated when the material is also subjected to heat for long periods of time.

Stress

This is created when a load is applied to a material. For example, a girder carrying the weight of a piece of machinery is under stress due to the weight of the machine. Where heavy weights or loads are involved, there may be some visible deflection or movement, but this may be part of the design.

Stress can occur during normal operations. For example, when components are subjected to warming and/or cooling, this will lead to those components expanding or contracting.

Stress corrosion cracking

This is where a material is subjected to both stress and corrosion. The corrosion has the effect of reducing the threshold of the material at a particular point (i.e. it weakens it), resulting in the stress causing the material to crack at that point. If the material had been subject to stress without corrosion, the crack would not have occurred. If the material had been subject to corrosion without stress, again, the crack would not have occurred. It is when these two factors combine that stress corrosion cracking can occur.

Stress Corrosion Cracking (SCC) can be highly chemical-specific, and certain pairings of materials with specific environments should be avoided. These include:

▷ Brass paired with ammonia
▷ Stainless steel paired with chlorides
▷ High strength steel paired with hydrogen

The following controls will help in the prevention of stress corrosion cracking:

▷ Selection of appropriate material
▷ Controlling service stress
▷ Use of corrosion inhibitors
▷ Coating material
▷ Isolating material from the local environment

Thermal shock

This is the stress introduced into a material as a result of a sudden and dramatic change in temperature.

Means of reducing thermal shock would be to:

▷ Change the temperature more slowly
▷ Change the temperature more evenly
▷ Increase the thermal conductivity of the material
▷ Reduce the material's coefficient of thermal expansion
▷ Increase the material's strength

Brittle fracture

If someone tried to bend an engineering file, which is made of a very hard but brittle metal, it would snap. This is an example of brittle fracture.

The conditions under which low- and medium-carbon steel can be affected by brittle fracture are as follows:

▷ A stress concentration must already be present in the steel. This may be a welding defect, a fatigue crack, a stress corrosion crack or a designed notch such as a sharp corner, a threaded hole or something similar.
▷ A tensile stress must be present. The tensile stress must be of sufficient magnitude to be able to cause deformation.
▷ The temperature must be relatively low for the steel concerned.
▷ Susceptible steel has been used.

The absence of any one of the above conditions will reduce the possibility of brittle fracture occurring.

What is meant by a safe operating envelope

A safe operating envelope is defined as the parameters and conditions a plant must operate within to ensure it is not subjected to excessive stress which might introduce or encourage failure modes. Some examples of parameters and conditions include:

▷ Safe working loads
▷ Maximum flow rates
▷ Maximum pressures
▷ Maximum and/or minimum temperatures

Use of knowledge of failure modes in initial design, process and safe operating procedure

When plant is being designed, careful thought and consideration should be given to the process itself, and the extent of stresses that may be encountered in extreme circumstances and how these stresses might be controlled.

Some simple control techniques include:

- ▶ Fitting expansion loops and expansion bellows in steam pipes.
- ▶ Ensuring the materials used are correct and capable of coping with the conditions they will be exposed to.
- ▶ Ensuring the thicknesses of the materials used are adequate to cope with the stresses they will be exposed to.
- ▶ Ensuring the strength of materials used is adequate.
- ▶ Ensuring equipment is correctly supported.
- ▶ Ensuring pipework and machinery are correctly aligned.
- ▶ Ensuring automatic shutdown trips are set to activate where parameters are exceeded.
- ▶ Ensuring the control systems are adequate.
- ▶ Ensuring operating procedures are always to hand.

Failure of the annular rim (bottom rim of storage tank)

The bottom plate in a storage tank is known as the annular plate and the junction between this bottom plate and the wall of the storage tank is known as the annular rim. The annular plate usually sits on a foundation of hardcore or a concrete ring wall, and is joined to the walls of the tank. Although the annular plate is not subject to high levels of stress, the joint of the plate to the walls is. This is because the weight of the product within the tank wants to push the walls outwards whilst at the same time push the annular plate downwards. This creates a high level of bending stress. The quality of the foundations will have a bearing on the downward deflection of the annular plate.

A further complicating factor to this stress is the fact that annular plates are prone to corrosion attacks both on the outer side where the tank shell sits on the annular rim, and on the underside of the annular plate where trapped water may lie undetected. This corrosion, coupled with the prolonged stress, can lead to stress corrosion cracking and failure occurring without warning.

Control measures should include ensuring the annular plate is kept as dry as possible by keeping the bunded area, where the tank sits, well drained. Other control measures include regular inspections of the annular rim and the application of effective maintenance.

Sub-element 3.1: Failure modes

Learning outcome

Explain the purpose of, and procedures for, investigating incidents and how the lessons learnt can be used to improve health and safety in the oil and gas industries.

 Revision exercise

Write your answers to the questions below on a separate sheet of paper without referring to the information in this book in the first instance. Once you have answered all of the questions, you can refer back to the revision guide to compare your answers and this will give you an indication of how much knowledge you have been able to absorb or whether you need to revise this section further.

Q1 **Describe** the following words or phrases in relation to materials:
(a) tension
(b) compression
(c) shear

Q2 **Describe** the following words or phrases in relation to materials:
(a) ductile
(b) malleable
(c) brittleness
(d) elasticity

Q3 **Describe** the following words or phrases in relation to materials:
(a) creep
(b) stress

    (c) stress corrosion cracking

    (d) thermal shock

    (e) brittle fracture

Q4 **Explain** what is meant by a 'safe working envelope'.

Q5 **Outline FOUR** measures that can be applied when designing a plant to control stresses in material.

Q6 **Outline** what stresses and conditions might cause the failure of the annular rim of a storage tank.

Sub-element 3.2: Other types of failure

Weld failures – the need for regular weld inspection and non-destructive inspection techniques

Introduction to weld failures

When a weld is being formed, certain chemical and metallurgical actions take place in the metal adjacent to where the weld is being deposited. Most welds that fail can be attributed to:

- Improper design of the weld joint
- Poor selection of base materials and filler materials
- Inappropriate welding processes
- Residual stresses
- Ineffective or non-existent inspection procedures
- Welded components operating outside their safe parameters

Types of weld failures – cracks

These are a significant element in weld failures. This is because the crack has the potential to grow into a total failure once the weld is under load conditions.

Hot cracks and cold cracks

Cracks in welds can be classified into two types: hot cracks and cold cracks. This refers to the condition of the weld when the crack first presents itself.

Hot cracks develop when the weld is still at a high temperature. Initially, they propagate between the grains of the weld material whilst it is still molten and develop into a full crack during the solidification process.

Cold cracks develop at low temperatures, generally because of stresses which develop as the weld solidifies. The influx of hydrogen during the weld process can also have an embrittling effect (loss of ductility) which can encourage cold cracking.

Porosity

This describes the presence of air bubbles within a weld. Porosity happens when the molten weld pool absorbs certain gases (nitrogen, oxygen or hydrogen) which become trapped and result in tiny bubbles once the weld solidifies. The main reason for this absorption is poor gas shielding during the welding process. Contamination from hydrogen – which stems from moisture – can be attributed to moisture being present on the electrodes, the fluxes or the components being welded.

Weld profile

This refers to the size of the weld. If a weld is too small in section then its load-carrying capability will be limited and may well not meet the parameters expected of the finished component. However, if a weld is too big this can also create problems which may lead to a weld failure. Too much root penetration, which is associated with welds regarded as too big, can create defects and cracks.

Weld testing techniques

The use of non-destructive testing has proved itself to be a valued procedure in reducing the number of weld failures occurring on components which are in service.

Weld testing techniques – magnetic particle inspection

This is where a magnetic field is introduced into the material to be tested. Ferrous iron particles coated in fluorescent dye are then applied, and any defects will attract these particles to the location. These areas will be highlighted when an ultraviolet light is shone onto the area being tested.

Weld testing techniques – dye penetrant inspection

This can detect surface-breaking defects in non-porous materials. The principle behind the process is that dye will penetrate into any surface-breaking defect and highlight it.

Weld testing techniques – ultrasonic flaw detection

This uses energy waves (high frequency vibrations – ultrasound) to read the exact structure of a component and thus detect any flaws or defects that may be present. The readings from the test procedure are displayed in a spiked graph form.

Weld testing techniques – radiography

This uses short wavelength electromagnetic radiation to penetrate the material being inspected and highlight any defects. This electromagnetic radiation is emitted from one side of the component and detected and measured on the opposite side of the component. This allows an analysis of the composition of the material to be made. This is usually in the form of a film or negative (X-ray).

Sub-element 3.2: Other types of failure

Learning outcome

Outline types of failure that may lead to loss of containment by hydrocarbons.

Revision exercise

Write your answers to the questions below on a separate sheet of paper without referring to the information in this book in the first instance. Once you have answered all of the questions, you can refer back to the revision guide to compare your answers and this will give you an indication of how much knowledge you have been able to absorb or whether you need to revise this section further.

Q1 **Give FOUR** causes of weld failure.

Q2 **Explain** what porosity is in relation to welds and how it occurs.

Q3 **Describe** the following non-destructive weld testing techniques:
   (a) magnetic particle inspection
   (b) dye penetration inspection
   (c) ultrasonic flaw detection
   (d) radiography

Sub-element 3.3:
Safety critical equipment controls

Learning outcome

Outline the controls available to maintain safety critical equipment.

🔑 **Key revision points**

Emergency shutdown (ESD) equipment and systems ☐

Safety Integrity Levels (SIL) for instrumentation ☐

Procedures for bypassing Emergency Shutdown (ESD) systems ☐

🔑 **Key revision points (continued)**

Blow down facilities, flare types ☐

Closed and open drain headers, sewers, interceptors ☐

Emergency Shutdown (ESD) equipment and systems

Emergency shutdown systems

A typical emergency shutdown system is made up of:

▷ Various sensors to detect any fire or escape of gas or vapour
▷ Valves and trip relays to isolate sections of the process
▷ A system logic for processing any incoming signals
▷ An alarm system to warn operators and control room staff of a potential adverse occurrence

The system processes the incoming signals and activates output commands in accordance with the cause and effect chart which will have previously been defined for that particular refinery or installation.

Emergency shutdown systems – typical actions

▷ Shutdown of part systems and equipment
▷ Isolate hydrocarbon inventories
▷ Isolate electrical equipment
▷ Stop hydrocarbon flow
▷ Depressurize/blow down
▷ Activate fire-fighting controls (water deluge, inert gas, foam system, water mist)
▷ Activate emergency ventilation control
▷ Close watertight doors and fire doors

Emergency shutdown systems – the voting system

To ensure any emergency shutdown system has the highest integrity level, the system is split and usually uses a triplicated microprocessor logic system. This triplicated system ensures that if one microprocessor fails, it does not compromise the implementation of the emergency shutdown system if required.

The system is based on an analysis of the information by voting on the inputs of signals received from fire and gas detectors positioned throughout the facility.

Actions taken will depend on a voting system of input signals from the detectors. One vote out of three will raise an alarm which will be investigated. Two votes out of three will activate the Emergency Shutdown (ESD) system.

Components of an emergency shutdown system

Components of an emergency shutdown system – Remotely Operated Shut-Off Valves (ROSOVs) and Emergency Shutdown Valves (ESDVs)

The principle of these valves is to isolate sections of the process or inventory quickly in order to reduce the amount of hydrocarbon available to feed a fire or leak.

Components of an emergency shutdown system – High Integrity Pressure Protection System (HIPPS) valves

When triggered, these valves isolate the process plant from the sources of high pressure before the safety parameters of the plant are exceeded, thus preventing loss of containment through a potential rupture.

113

Components of an emergency shutdown system – deluge systems

Water deluge systems are a means of fighting a fire with large amounts of water. They are generally positioned in areas where hydrocarbons are processed or stored, as well as in areas where there is the potential for an uncontrolled release of gas which could result in a fire or explosion.

Components of an emergency shutdown system – fire and gas detection systems

These are made up of a number of detectors, each of which is specifically selected and strategically positioned so as to raise awareness of an adverse activity at the earliest opportunity so that appropriate action can be taken.

Components of an emergency shutdown system – vent and blow down system

This is a system that is used to relieve pressure within a process and deal with it in a safe and controlled manner. The pressure relief may involve liquid or gas, depending on the process and/or the situation.

Level of shutdown

The level of shutdown needed in response to inputs from an adverse situation needs to reflect the severity of the situation. However, the requirements of each plant, installation or platform will vary accordingly and the shutdown level hierarchy will reflect this

Table 13.1 – Offshore platform shutdown level response hierarchy

Shutdown level	Cause	Effect
Level 1	Failure of any non-critical equipment	Affected equipment shutdown and standby equipment started. Minimal effect on production.
Level 2	Failure of any critical equipment	Production terminated whilst equipment is replaced. Sections of system are isolated to allow

		equipment to be replaced. Blow down and venting may be required.
Level 3	Fire or gas alarm activated	Production may be affected depending upon the location and extent of hazard. Equipment may be shut down or isolated depending upon which zone is affected and the sensitivity level in that zone.
Level 4	Manual initiation of emergency shutdown	Total production shutdown as well as non-essential utilities being shut down. Product flow from wellhead isolated. Prepare to Abandon Platform Alarm (PAPA) initiated to muster personnel to lifeboat stations.
Level 5	Manual initiation of abandonment	Platform abandon status adopted, inventory vented, Subsea Isolation Valves (SSIVs) shut and all essential utilities shut down.

Source: Adapted from www.hse.gov.uk/comah/sragtech/techmeascontsyst.htm; www.hse.gov.uk/pubns/chis2.pdf

Table 13.2 – Onshore process plant shutdown level response hierarchy

Shutdown level	Type of shutdown	Response
Level 1	Unit shutdown	This level involves the shutdown of single process or function. It can be activated automatically by sensors or by manual means. Its aim is to prevent the equipment from operating outside its safe operating envelope.
Level 2	Process shutdown	This level involves shutting down all process systems. It is activated automatically by sensors. Its aim is to isolate all process equipment and limit the potential of an escalating emergency situation developing.
Level 3	Emergency shutdown	This level involves shutting down and isolating specific process equipment through the activation of emergency shutdown valves. It can be activated automatically, or by manual means.

Shutdown level	Type of shutdown	Response
		Its aim is to limit the consequences during an emergency situation.
Level 4	Emergency depressurization shutdown	This level involves shutting down, isolating and depressurizing equipment and processes by way of opening valves to allow blow down and venting to occur. It is activated manually. Its aim is to reduce the potential for overpressure and to release hazardous vapours and gases from the process.

Source: Adapted from www.hse.gov.uk/comah/sragtech/techmeascontsyst.htm; www.hse.gov.uk/pubns/chis2.pdf

Safety Integrity Levels (SILs) for instrumentation

Safety instrumented systems

A safety instrumented system can be defined as 'a system made up of sensors, logic solvers and actuators which has the ability to take a process to a state of safety when predetermined parameters are exceeded or safe operating conditions are breached'.

An example of this might be the sensor (e.g. gas or flame sensor), the controller (the computer logic solver) and the actuator (the electrical device which activates the emergency shutdown device). All of these individual components, and the system as a whole, need to have a level of integrity or dependency which is in line with the consequences of failure. The greater the consequence, the higher level of integrity or dependency needed.

Safety Integrity Level (SIL)

There are four discrete safety integrity levels: SIL 1–SIL 4. The higher the SIL number, the higher the associated safety level requirement is, and this needs to be coupled with a lowering of the probability that the safety instrumented system will fail. The probability of failure is defined numerically below.

116

▷ SIL 1 – This level of integrity is where the acceptable probability of failure is between 1 in 10 occasions, and 1 in 100 occasions. It is required where the potential for relatively minor incidents is involved with limited consequential outcomes.

▷ SIL 2 – This level of integrity is where the acceptable probability of failure is between 1 in 100, and 1 in 1,000 occasions. It is required where the potential for more serious, but limited incidents is involved and where the consequences may result in serious injury or death to one or more persons.

▷ SIL 3 – This level of integrity is where the acceptable probability of failure is between 1 in 1,000, and 1 in 10,000 occasions. It is required where the potential for serious incidents is involved and where the consequences may involve a number of fatalities and/or serious injuries.

▷ SIL 4 – This level of integrity is where the acceptable probability of failure is between 1 in 10,000 occasions, and 1 in 100,000 occasions. It is required where the potential for a catastrophic incident exists.

The required safety integrity level for each safety instrumented system is established by conducting a hazard and risk assessment (a hazard and operability study or hazard identification study). This will determine the consequences of the safety instrumented system not functioning as required, which in turn will indicate the appropriate safety integrity level for that safety instrumented system.

Procedures for bypassing Emergency Shutdown (ESD) systems

Bypassing ESD

All emergency shutdown systems and fire and gas systems need to be tested, inspected and/or maintained on a regular basis to ensure they are functioning as required. These testing and/or maintenance procedures involve the temporary bypassing of safety system interlocks. These are the devices within the logic system which activate the alarm as well as sending a signal to the actuator on the emergency shutdown component itself.

117

All inhibits or overrides should be logged and a record kept. An inhibit log should record the following information:

▶ The safety function being inhibited
▶ The time and date the inhibit was applied
▶ A cross reference to the relevant permit-to-work or protective systems isolation certificate where applicable
▶ The time and date of each reassessment
▶ The time and date the inhibit was removed

Other control measures to be applied to inhibits and overrides include:

▶ They should only be applied with the authority of a senior person or manager.
▶ They should be time bound at the outset. Equipment must be shut down if the emergency shutdown system is not reinstated within the designated time.
▶ Their implementation should be communicated to all operations personnel as well as adjacent plants.

Blow down facilities, flare types

The term 'blow down' refers to the action of venting gas or relieving pressure from a process or production system

Types of flare

Flares work on the basis of introducing air at the point where fuel gas is combusted so that efficient and complete combustion is achieved and smoke is kept to a minimum.

Steam assisted flares employ a system which injects steam into the flare at the flare tip. This introduces air into the flame, which makes the flame more efficient whilst at the same time reducing the tendency for the flame to emit smoke.

Air assisted flares use a fan to direct air flow to the flare tip. This air flow assists in two ways. First, it adds momentum to emission of the fuel gas; second, it provides a contribution towards the air needed for efficient combustion.

Unassisted flares use only the pressure of the emission of the fuel gas to provide momentum for the flame. The pressure of the fuel gas helps mix the gas with air for combustion.

Multi-point pressure assisted flares use the principle of high exit velocity of the fuel gas from each burner. This induces the mixing of fuel gas with air for efficient combustion. Generally each burner is approximately 2.5 metres tall and they are arranged in rows, the formations of which are known as Multipoint Ground Flares (MPGF). The area they cover can be as big as a football field.

Closed and open drain headers, sewers, interceptors

Drainage systems

Drainage systems are a means of safely collecting any residual process fluids, hydrocarbon liquid and/or chemical spills, deluge water and rain water, and transporting it to an appropriate location where it can be dealt with or disposed of appropriately.

Drainage – open drain systems

A drainage system that collects fluids that spill onto the ground is called an 'atmospheric drain', 'gravity drain' or 'open drain'. This system will consist of collecting funnels or trays known as 'tundishes' which are strategically positioned so as to channel any liquid they collect through a series of drain pipes to an open drain header before being routed to a slop tank ready to be dealt with appropriately. There will be separate drainage systems to deal with 'safe areas' and 'hazardous areas'.

Drainage – closed drain systems

A drainage system that is required to be connected to a pressure vessel in order to drain off fluids is called a 'pressure drain system' or 'closed drain system'. Whenever a vessel is drained of liquid under pressure into a pressure or closed drain system, it must be assumed that the liquid contains a certain amount of dissolved gases. Furthermore, the flow of liquid from the pressurized vessel will be followed by a certain amount of gas (known as gas blow by). This gas, in both its forms, can represent a hazard if it is not dealt with appropriately, for example by using a blow down facility so that gas is vented away to a flare.

Drainage – interceptors

Drainage interceptors are a means of collecting contaminated water before it is discharged to a foul drain or surface drain. Typically, interceptors have three separate chambers, with the divisions between chambers extending down to the bottom, and low level pipes connecting the chambers. This is so that when the contaminated water enters the first chamber it can separate (oil will naturally float on top of water) and be extracted. The water is then directed to the second and third chambers via the low level pipe, where any residual oil is also allowed to separate and be extracted. Finally, the water from chamber three is channelled into either a foul or surface drain, whichever is appropriate.

Sub-element 3.3: Safety critical equipment controls

Learning outcome

Outline the controls available to maintain safety critical equipment.

 Revision exercise

Write your answers to the questions below on a separate sheet of paper without referring to the information in this book in the first instance. Once you have answered all of the questions, you can refer back to the revision guide to compare your answers and this will give you an indication of how much knowledge you have been able to absorb or whether you need to revise this section further.

Q1 **Identify** the general groups of components of an emergency shutdown system.

Q2 **Give FIVE** typical actions an emergency shutdown system might perform.

Q3 **Explain** what the 'voting system' is within the response section of an emergency shutdown system.

Q4 **Explain** what 'safety integrity levels' are.

Q5 **Explain** under what circumstances an emergency shutdown system might need to be bypassed.

Q6 When an emergency shutdown system is bypassed this should be recorded in an inhibit log. **Identify FIVE** pieces of information relating to the bypass that should be recorded in the log.

Q7 **Describe** what the term 'blow down' refers to.

Q8 **Describe** what an 'open drain system' is and its function.

Q9 **Describe** what a 'closed drain system' is and its function.

Q10 **Describe** what an 'interceptor' is and its function.

Sub-element 3.4:
Safe containment
of hydrocarbons

Learning outcome

Outline the hazards, risks and controls available for safe containment
of hydrocarbons offshore and onshore.

🔑 Key revision points

Hazards and risks including overfilling, effects of vacuum,
overloading of foundations and failure modes for tank shells and
associated pipework ☐

Floating roof tanks, landing the roof, sinking the roof and rim
seal fires/failures ☐

Fixed roof storage tanks, pressure and vacuum hazards ☐

Key revision points (continued)

Bunding of storage tanks including volume and area sizing, construction and valving arrangements ☐

Filling of tanks, overfilling/alarms/tanker connections ☐

Pressurized/refrigerated vessels for Liquefied Petroleum Gas (LPG), Liquefied Natural Gas (LNG) and carbon dioxide (CO_2) ☐

Loss of containment and consequences ☐

Decommissioning of plant and associated facilities ☐

Management of simultaneous operations ☐

Hazards and risks including overfilling, effects of vacuum, overloading of foundations and failure modes for tank shells and associated pipework

Storage tanks

Hydrocarbons are usually stored in tanks. The type and quantity of the product being stored will dictate the type of tank used as well as the measures used to control the risk of it failing.

Storage tanks – capacity

Each tank will have a prescribed maximum volume, although the density of oil varies from approximately 0.8 tonnes per m^3 to 1.3 tonnes per m^3 depending on its type and grade, and the prescribed volume may have to be reduced accordingly for that reason.

Storage tanks – integrity management

The following management practices are regarded as good practice with regard to managing storage tank integrity.

124

▷ Tanks containing hazardous substances should be identified and entered in the plant register.

▷ Operators should maintain tank data files.

▷ Compatibility assessments should be undertaken (e.g. tanks on multi-product service).

▷ Tanks should be subject to formal periodic maintenance and inspection.

▷ Inspection and maintenance of tanks should only be carried out by persons who are experienced, qualified and competent.

▷ Schemes of inspection should be established and agreed between operators and competent persons.

▷ Appropriate inspection techniques should be utilized, depending on deterioration mechanisms.

▷ Inspection reports and checklists should be of high quality.

▷ Where necessary, recommendations from inspection reports should be actioned promptly.

▷ Assessment for fitness for service should be carried out following tank inspection and significant changes to process or operating conditions.

▷ Operators and competent persons should have knowledge of, and adopt the recommendations given in, the relevant guides, codes and standards.

▷ Tank examination schemes should include both internal and external inspections.

Evolving damage mechanisms

There are many types of evolving damage mechanisms which can affect storage tanks, many of which can work simultaneously. They include the following:

Evolving damage mechanisms – corrosion

This is the gradual deterioration of a substance by way of a chemical reaction with its immediate environment. Most storage tanks are made from carbon steel, which makes corrosion a primary cause of

125

deterioration. Furthermore, corrosion can occur both internally and externally, which makes the monitoring of corrosion a difficult issue.

Evolving damage mechanisms – erosion

This is the process of material being worn away by the constant movement of product flowing over the surface. Areas such as filling and discharge points which experience large amounts of product flow are the most vulnerable points. The main control measure for erosion is to increase the thickness of the material where erosion is identified as a threat.

Evolving damage mechanisms – creep associated with thermoplastic tanks

This is where the tank material stretches over a period of time when it is under stress from the weight of the product being stored. However, this condition only affects non-metallic thermoplastic tanks. Increases in temperature tend to compound the problem, with High-Density Polyethylene (HDPE) being particularly vulnerable and losing its strength as the temperature increases.

Evolving damage mechanisms – fatigue

Pressure vessel storage tanks are more prone to fatigue defects than tank storage vessels. This is because the membrane stresses in pressure vessels are much greater.

For tank storage vessels which are subject to a cycle of frequent filling and emptying, certain parts of the tank's construction can be vulnerable, particularly welded joints.

Control measures revolve mainly around inspection and monitoring so that any problems can be detected and dealt with before they become an issue.

Evolving damage mechanisms – chemical attack

The most common materials used in the construction of storage tanks are carbon steel, stainless steel and Glass Reinforced Plastic (GRP).

Carbon steel is suitable for a large number of products, although some products with an acid content may react with carbon steel.

Stainless steel is used where purity of the stored product is paramount. However, welded areas in stainless steel tanks can be prone to stress corrosion cracking.

Glass Reinforced Plastic (GRP) suffers from ultraviolet light degradation over time, and has a reduced ability to perform as product temperature is increased.

Evolving damage mechanisms – brittle fracture

This tends to be most prevalent either during hydro-testing shortly after construction has been completed or during the first filling in cold weather.

Mechanical damage

Mechanical damage – impact

An impact on a metal tank will have a distorting effect which can compromise tanks with floating roofs. Non-metallic tanks have poor impact resistance and may well fracture if the impact is serious.

Control measures include placing barriers in areas where impacts may occur. Similarly, bunding may act as a protective barrier. Good inspection regimes will also pick up on externally inflicted damage.

Mechanical damage – settlement which is non-uniform

Subsidence can be caused by constant weight on weak or compressible terrain. The weakness or compressibility may not be uniform, which can lead to distortions in the tank structure. Foundations which are designed and built to compensate for local conditions are essential.

Two further causes of foundation movement are frost heave, where the ground is subject to frequent freezing and thawing, and ground movement caused by high tides in areas close to the sea.

127

A precautionary measure where these factors are recognized as issues is to fit connecting pipes with flexible joints or bellows to allow for movement.

Mechanical damage – over-pressurization

This is where pressure is allowed to build up to unacceptable levels and can lead to a tank rupture. This is most likely to happen when the tank is being filled and the pressure relief valve fails. Controls include regular inspection and maintenance of these valves and any sensors associated with them.

Mechanical damage – vacuum

This generally affects fixed roof storage tanks. As the tank is gradually emptied of product, if the relief valve or vent becomes blocked or fails, a vacuum will build up and cause the tank to distort. Again, controls include regular inspection and maintenance of these valves and vents, together with any sensors associated with them.

Mechanical damage – excessive external loads

The most common external loads which can cause damage to a storage tank are snow and ice. These can cause buckling where the loads are excessive and should be cleared before they cause problems. Rainwater can also be an issue, especially on floating roofs where drains have become blocked.

Mechanical damage – wind loads

These can cause tanks to lift, especially if they are not anchored down or have little product in them.

Mechanical damage – tanks floating off their foundations

In a situation where rainwater is allowed to build up inside a bunded area with a tank in it which is not holding a lot of product and which has not been anchored down, it is possible for the tank to float off its foundations.

Controls are to keep bunded areas free of rainwater by pumping it out on a regular basis.

Floating roof tanks, landing the roof, sinking the roof and rim seal fires/failures

External floating roof tanks

An external floating roof tank is made up of a cylindrical shell with a fixed base. This shell has a roof within it which floats up and down on top of any liquid stored within the tank (Figure 14.1). This means that there is no vapour space – known as ullage – in a floating roof tank. This lack of ullage in a floating roof tank reduces product loss due to evaporation. A rim seal between the roof and the inner tank wall also ensures evaporation is kept to a minimum.

The main disadvantage of an external floating roof is that snow can accumulate on the roof and build up to such a level as to cause the roof to sink. The same can happen with rainwater if the drain from the roof becomes blocked.

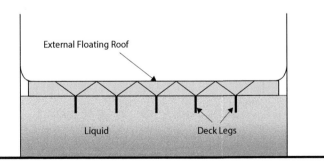

Figure 14.1 Section through external roof storage tank

Source: Wise Global Training.

Internal floating roof tanks

An internal floating roof tank uses the same principle as an external floating roof, but in addition the tank has an external, fixed roof (Figure 14.2).

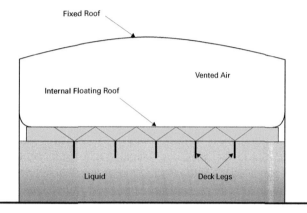

Figure 14.2 Section through internal roof storage tank

Source: Wise Global Training.

Features of internal floating roof tanks are as follows:

▶ The internal floating roof tends to be of very lightweight construction and, as such, is not suitable for people to walk on.
▶ Internal floating roof tanks are free venting. That is to say, they are not fitted with Pressure and Vacuum (P&V) valves.
▶ Internal floating roof tanks tend to be used for storing material with a low flash point, such as gasoline.

Issues with floating roofs

Issues with floating roofs – landing the roof

This is where the liquid in the tank falls far enough for the legs on the underside of the floating roof to land on the base of the tank. The void between the liquid and the roof will grow, which will allow a build-up of vapour to occur. This has the potential to cause a fire and/or an explosion.

Issues with floating roofs – sinking the roof

This can be caused when the roof becomes unbalanced for some reason and sinks into the liquid. Reasons for becoming unbalanced include:

▸ Build-up of rain or snow
▸ Use of access ladders on one side of the roof
▸ In certain parts of the world, earthquake activity may have a destabilizing effect on the roof

With internal roofs, if the roof sinks the ullage space will quickly fill with flammable vapours and create a fire/explosion hazard. However, the volatile atmosphere will be contained within the tank beneath the fixed roof.

With external roofs, if the roof sinks, the flammable vapours are free to evolve and drift away and may meet a source of ignition.

Issues with floating roofs – rim seal failures

Floating roofs move up and down within the shell of the tank according to the level of liquid in the tank. The space between the tank wall and the roof is kept airtight by a rim seal which ensures vapours don't escape and contaminates, such as rainwater, don't get into the stored liquid.

Most systems have a double seal so that there is a back-up in place should the primary seal fail. Furthermore, there is usually a detection system built onto the roof so that if a leak does occur, it can be detected and dealt with. The detection system may well also incorporate an automatic rim seal fire-fighting system.

Without these safety features, vapours could escape unnoticed. In such cases there have been occasions where an electrical storm has created a set of circumstances which have created a spark adjacent to the tank. This has subsequently ignited the vapour and caused a fire.

Fixed roof storage tanks: pressure and vacuum hazards

When liquid is pumped into a storage tank the atmosphere in the void needs to be vented out. If this does not happen the pressure will become unsustainable and the tank may rupture.

The contra view is also true, in that when a tank is emptied the liquid reduces and the void above the liquid needs to have air vented in so that a vacuum is not created. If the air inlet valve is faulty, and a vacuum is created, this may well cause the tank to collapse.

There are other factors which can cause increased or decreased pressure within the tank. These include:

▷ Storing a volatile product will cause gases to evolve and increase pressure.
▷ Warm weather or direct sun on the tank will warm the product up and make it expand, thus increasing pressure.
▷ Cold weather will cool the product down and cause it to contract, thus decreasing pressure.

Normally, tanks have a Pressure and Vacuum (P&V) relief valve fitted on or near the top of the tank. This allows vapour or gas to escape or, in the case of a reduction in pressure within the tank, allows air to enter the tank. This ensures the integrity of the tank is retained.

Consequently, the P&V relief valve is a critical piece of equipment and the integrity of the tank is dependent on it working properly. It should therefore be regularly inspected and maintained.

Another method of controlling changes in pressure is to employ an expansion or pressure exchange system. This is where the expelled vapour or gas is directed to another tank rather than to atmosphere. Where a vacuum has to be dealt with, the system draws in an inert gas from an external source.

This system ensures that volatile gases or vapours are not released to atmosphere in an uncontrolled fashion.

Bunding of storage tanks, including volume and area sizing, construction and valving arrangements

Bunding of tanks

Bunds are designed to contain any spillages and stop them escaping; this includes seeping into the ground. Consequently, bunds should be built on an impervious base and the material the bund is constructed from should also be impervious.

A bund needs to contain any amount of spillage from the tank(s) it surrounds. Therefore the capacity of the bund must be equal to the total maximum content of the tank(s) within the bund plus an extra 10 per cent.

The bund itself must be complete, that is to say it must not be breached. Any pipelines must go over the bund rather than through it. Access to the tanks inside the bund must again be over the bund rather than through a gate.

Tanks can be individually bunded, or bunded as a group. However, where tanks are bunded as a group, it is important to ensure that products within the tank grouping will not conflict with one another if they leak at the same time. Finally, the bund wall should be substantial enough to maintain its integrity if a fire occurs.

Filling of tanks, overfilling/alarms/tanker connections

Transfer of material between road/rail tankers and tanks

The safe transfer between road/rail tankers and tanks should include:

▶ Securing the road/rail tanker – brakes applied and engine turned off.
▶ Ensuring that hoses are suitable for the product being discharged and the operating pressures they will be subjected to.
▶ Ensuring the connections of pipes and hoses to be used in the transfer operation are secure.

133

▷ Positioning of drip trays beneath all connections and ensuring there is close monitoring of connections during transfer operations.

▷ Making a good bonding connection between the road/rail tanker and the loading/unloading equipment. This will ensure that both the tanker and loading equipment have the same electrical potential.

▷ Making ready fire control measures, such as fire-fighting equipment, before transfer commences.

▷ Noting emergency escape routes before transfer commences.

▷ Closely monitoring the flow rates and fill capacities. This will include alarm systems to indicate when tanks are nearing their filling point.

▷ Having adequate venting arrangements in place to ensure vapour is dispersed properly and safely. This will include monitoring wind direction and strength.

▷ Low wind speed can be an added hazard, as the dispersion of vapour in these conditions is minimal and it can build up in dangerous quantities without being apparent. If there is an ignition source nearby, the results can be catastrophic.

▷ Having venting arrangements to the recipient tank in place as well as similar arrangements for the donor tank, as air, or more likely inert gas, will have to replace the volume of product transferred.

Pressurized/refrigerated vessels for Liquefied Petroleum Gas (LPG), Liquefied Natural Gas (LNG) and carbon dioxide (CO_2)

Liquefied Petroleum Gas (LPG) storage

To store LPG effectively, it has to be converted from a gas to a liquid, which means it is stored at a temperature of between 0°C and −44°C.

Liquefied petroleum gas containers that are subjected to fire of sufficient duration and intensity can undergo a Boiling Liquid Expanding Vapour Explosion (BLEVE). Consequently, LPG spheres should be protected by a fire water deluge system, and this will have to be supplemented with an adequate fire water runoff system.

Large, spherical LPG containers may have wall thicknesses of up to 15 cm. Furthermore, they will ordinarily be equipped with an approved pressure relief valve.

134

Liquefied Natural Gas (LNG) storage

Liquefied Natural Gas (LNG) needs to be stored at −162°C in order to maintain its liquid state. This effectively reduces its volume 600-fold.

Storage tanks that are designed to store LNG have double skins, with the inner skin containing the LNG and the outer skin containing insulation materials.

Carbon dioxide (CO_2) storage

Liquid carbon dioxide (CO_2) storage tanks usually have a storage capacity range from 3,000 kg to 50,000 kg. The vessels can be horizontal or vertically mounted.They have an inner steel tank cloaked by a layer of 175 mm thick insulation and are covered with aluminium cladding.

Loss of containment and consequences

Loss of containment is the one single event which has the greatest potential for catastrophic consequences, including loss of life and plant. Understanding what can happen to that product once it is released is therefore essential in implementing controls to mitigate the effects.

Jet fires

A jet fire is where a fire emanates from a source of fuel which is being continuously released in a particular direction with significant force. Jet fires reach full intensity almost instantly. However, the contrary is also true in that they can be turned off quickly if the source of fuel can be terminated.

Pool fires

A pool fire is a fire which burns above a horizontal pool of vaporizing hydrocarbon fuel. Pool fires can be static, where the pool of fuel remains in one position, or they can be running fires where the pool of fuel runs like a river.

How hydrocarbon vapour clouds are generated and their potential consequences

Any release of flammable liquids or gases can result in the formation of a cloud of flammable vapour which, as it is likely to be denser than the surrounding atmosphere, will stay at a low altitude as opposed to rising and dispersing in the upper atmosphere. If the cloud then meets a source of ignition, and it is within its flammability limits, it will result in a fire or explosion.

BLEVE – Boiling Liquid Expanding Vapour Explosion

Any vessel which is partly filled with pressurized hydrocarbon liquid (e.g. liquefied petroleum gas) will have a certain amount of space above it filled with vapour. If the vessel is subjected to a fire, the pressure in the tank will increase due to the liquid going above its boiling point and turning into a vapour.

The pressure relief valve will allow the overpressure to be vented to atmosphere in the first instance, but this will reduce the amount of liquid in the tank still further and the potential for the flame to engage with a section of the tank containing vapour and not liquid will increase. If this happens, the tank wall will weaken at this point as the heat transfer to vapour is much less efficient than to a liquid.

The result is likely to be a sudden and catastrophic failure of the vessel, with a discharge of vapour followed by an explosion when it reaches the flames. This, in effect, is a Boiling Liquid Expanding Vapour Explosion or BLEVE.

Confined Vapour Cloud Explosion (CVCE)

A confined vapour cloud explosion is an explosion following a leak of vapour which occurs in a confined space, such as a building or a tank.

Unconfined Vapour Cloud Explosion (UVCE)

An unconfined vapour cloud explosion is an explosion following a leak of vapour which occurs in an unconfined space (outdoors).

136

Pipelines, their protection, surveying, maintenance, security against arson and illegal tapping

Because of the high volumes of product involved, and because once flow through a pipeline stops, it stops completely, pipelines need to be surveyed and maintained to the highest standards.

Surveys, inspections and maintenance activities can be carried out using Pipeline Inspection Gauges (PIGs). These devices are inserted into the pipeline at special insertion points and can perform various operations including inspections, cleaning and other maintenance operations. They can perform all these operations without having to stop the flow of product in the pipeline.

Pipelines can be damaged by trailing anchors under the sea and trawling activities. Deliberate damage can come from arson or terrorist attacks or from illegal tapping. Security devices are available which can monitor pipelines that are land-based or sub-aqua based. These work using fibre optics and can monitor both pipeline security and leak detection.

On land-based systems, they can monitor both underground and overground pipelines. As well as leaks, they will detect any activity being undertaken on the pipeline and relay its location back to a control room or security centre.

Sub-aqua systems are capable of detecting leaks and even diver activity within the vicinity of the pipeline.

Decommissioning of plant and associated facilities

There are five distinct stages in the decommissioning process:

1   All options for the physical dismantlement and/or removal of the installation are assessed and the best option is selected. This is then put through a detailed planning process involving all engineering, safety and environmental aspects.
2   Production/extraction is stopped and all the wells are plugged.

137

3   The installation is dismantled and/or removed in line with the
     agreed plan of action.
4   Those parts of the installation that have been removed are
     disposed of or recycled.
5   A sea bed survey is carried out to ensure nothing untoward
     has happened to the original location of the installation. If any
     part of the installation remains, ongoing monitoring will be
     implemented.

Management of simultaneous operations

Simultaneous operations (SIMOPS) or activities can be described
as a period in time when more than one operation or activity is
being conducted at the same time, with the possibility that these
activities could clash and bring about an undesired event or set of
circumstances.

It is important that all work being undertaken at the same time, or
which overlaps within a specific time frame, is identified before that
work commences. It is then possible to review all activities and see
where any conflict might occur.

The parties involved in simultaneous operations should meet
to discuss the extent of their work and their requirements. The
intention of the meeting is for each party to be able to draw up
a work-specific dossier which covers how their activities will be
conducted whilst taking into account all the other parties' activities
and avoiding any conflict. The meeting should resolve the following
issues:

▶ The extent of the responsibilities of each party
▶ Nomination of a responsible person for each party
▶ Ensuring that responsible persons know of each other and have
  contact arrangements in place
▶ Identifying the requirements of each party
▶ Identifying the time frame of the actual work activity of each
  party

Sub-element 3.4: Safe containment of hydrocarbons

Learning outcome

Outline the hazards, risks and controls available for safe containment of hydrocarbons offshore and onshore.

✎ Revision exercise

Write your answers to the questions below on a separate sheet of paper without referring to the information in this book in the first instance. Once you have answered all of the questions, you can refer back to the revision guide to compare your answers and this will give you an indication of how much knowledge you have been able to absorb or whether you need to revise this section further.

Q1   **Explain** why the maximum volume of product stored in a storage tank can vary.

Q2   **Outline SIX** management practices which can be regarded as good with regard to maintaining the integrity of storage tanks.

Q3   With regard to storage tanks, **explain** what the process of 'corrosion' is and where it can take place.

Q4   With regard to storage tanks, **explain** what the process of 'erosion' is and where it can take place.

Q5   With regard to thermoplastic storage tanks, **explain** what the process of 'creep' is.

Q6   With regard to storage tanks, **explain** what the process of 'fatigue' is and where it can take place.

Q7   With regard to storage tanks, **give TWO** examples of how 'settlement which is non-uniform' can happen.

Q8   **Explain** under what circumstances it is possible for a storage tank to be floated off its foundations.

Q9   **Explain** what 'landing the roof' of a storage tank is and the hazards associated with it.

Q10   With regard to external roof storage tanks, **give TWO** reasons why the floating roof of a storage tank might sink.

139

Q11 **Describe** what safety features can be incorporated into a floating roof storage tank which will address the issue of rim seal failure.

Q12 With regard to fixed roof storage tanks, **explain** how 'over-pressurization' can occur, its effects and the control measures required to prevent it.

Q13 With regard to fixed roof storage tanks, **explain** how 'vacuum' can occur, its effects and the control measures required to prevent it.

Q14 **Explain** what bunding of storage tanks is and its purpose.

Q15 With regard to the transfer of material between road/rail tankers and storage tanks, **give FIVE** safety measures which should be incorporated into the procedure.

Q16 **Describe** the characteristics of a 'jet fire'.

Q17 **Describe** the characteristics of a 'pool fire'.

Q18 **Describe** how a Boiling Liquid Expanding Vapour Explosion (BLEVE) might occur.

Q19 **Describe** what a Confined Vapour Cloud Explosion (CVCE) is.

Q20 **Describe** what a Unconfined Vapour Cloud Explosion (UVCE) is.

Q21 **Outline TWO** examples of how pipelines can be damaged.

Q22 **Outline** the **FIVE** stages of decommissioning.

Q23 **Outline** the measures which should be taken to address potential conflicts when simultaneous operations are undertaken.

Sub-element 3.5:
Fire hazards, risks and controls

Learning outcome

Outline the fire hazards, risks and controls relating to hydrocarbons.

🔑 Key revision points

Lightning ☐

Fire triangle and potential consequences of explosions and thermal radiation ☐

Electrostatic charges ☐

The identification of ignition sources ☐

Zoning/hazardous area classification and selection of suitable ignition-protected electrical and mechanical equipment and critical control equipment ☐

Lightning

Threats from a lightning strike include:

- Sparks which can cause a fire or explosion
- Power surges to electrical equipment, particularly monitoring and safety devices, which can render them inoperable

Protection from lightning strikes is a specialist area requiring expert knowledge as to what systems are suitable for each facility. In general, the necessary precautions are:

- Keep the lightning channelled far away from the immediate neighbourhood of flammable and explosive materials
- Avoid sparking or flashover in joints and clamps, and at nearby components
- Prevent the overheating of conductors
- Prevent flashover or sparking due to induced voltages
- Prevent raising the potential of the earth termination system
- All metal containers to be of sufficient thickness (usually 5 mm minimum)
- Down-conductors to be fitted to all other metal structures and in sufficient numbers as to subdivide any current surge adequately
- All earthing systems to be interconnected to a single earth termination system. This usually takes the form of a mesh or grid pattern around the site.

Fire triangle and potential consequences of explosions and thermal radiation

The fire triangle

Fire is made up of three interdependent elements known as the fire triangle. These are:

- Heat or a source of ignition
- Fuel
- Oxygen

A fire can be prevented or extinguished by removing any one of these elements.

Explosions

An explosion is a type of fire, but one which combusts with such a rapid force that it causes an effect known as overpressure (explosion).

There are three types of explosion that are associated with the oil and gas industry. These are:

1. Boiling Liquid Expanding Vapour Explosion (BLEVE)
2. Confined Vapour Cloud Explosion (CVCE)
3. Unconfined Vapour Cloud Explosion (UVCE)

Thermal radiation

This is the transfer of heat from one source to another. This can be a structure or a person. Where the recipient source is a person the initial effect of exposure to a source of heat (fire) is to warm the skin. This then becomes painful as the amount of energy absorbed increases. Thereafter, second degree burns begin to take effect, with the depth of burn increasing with time for a steady level of radiation. Ultimately, the full thickness of the skin will burn and the underlying flesh will start to be damaged, resulting in third degree burns.

When plant, including pipework and vessels, is exposed to thermal radiation the effect is the transfer of heat to the product inside the plant. This can change the characteristics of the product and make it less stable. These characteristics include the potential to make the product expand and/or increase the amount of vapour given off, amongst other things. This can result in loss of containment, with an ensuing vapour cloud explosion, jet fire, pool fire or running liquid fire.

Electrostatic charges

Whenever a liquid moves against a solid object, such as the inside of a pipe, it generates a static electrical charge. The most common cause of static electricity build-up is where there is a flow (transfer) or movement (mixing process) of liquid within a process.

143

The amount and rate of static generation can be dictated by a number of factors. These factors, or their elimination or reduction, can also be used to control the risks associated with static electrical generation. These include:

▶ The conductivity of the liquid
▶ The amount of turbulence in the liquid
▶ The amount of surface area contact between the liquid and other surfaces
▶ The velocity of the liquid
▶ The presence of impurities in the liquid
▶ The atmospheric conditions. Static build-up is enhanced when the air is dry.

Electrostatic charges – piping systems

The flow of liquid through piping systems can generate a static charge. Control measures include keeping the rate and velocity of the liquid low.

Electrostatic charges – filling operations

Filling operations which involve large flows of liquid and splashing generate turbulence. This turbulence allows the large amounts of liquid to pass against vessel surfaces which in turn generates a static charge. Control measures include:

▶ Ensuring filling operations do not involve the free-fall of liquids. This will reduce the amount of splashing taking place.
▶ Lowering the velocity of the liquid being filled.
▶ Ensuring fill pipes touch the bottom of the container being filled.
▶ Tanks which have been filled with products that have a low conductivity, i.e. jet fuels and diesels, should be given time to relax before the process continues.
▶ Tanks which have been filled with product should not have any ullage (vapour space) for a set period of time. Nor should any dipping of the product take place, again for a set period of time.

Electrostatic charges – filtration

Filters have large surface areas and this can generate as much as 200 times the amount of electrostatic charge in a piping system that has a filtration system within it as compared with the same piping system without filtration.

Control measures include ensuring good bonding and grounding is in place.

Electrostatic charges – other issues

▶ Liquids which have particles within them are more susceptible to the generation of static charge than those without.
▶ Static can be generated when liquids are mixed together.
▶ Piping or vessels which allow a space for vapour to accumulate are a particular concern, as any spark generated from a discharge of static electricity may cause an explosion inside the pipe.

Methods of controlling static charges

Although generating static electricity cannot be totally eliminated, the rate of generation and its accumulation can be reduced by the following control measures.

Methods of controlling static charges – additives

In some instances, anti-static additives can be introduced to reduce static charge build-up.

Methods of controlling static charges – bonding and grounding

A bonding system is where all the various pieces of equipment within a process system are connected together. This ensures that they all have the same electrical potential, which means there is no possibility of a discharge of electricity, by way of a spark, from one piece of equipment to another.

Grounding is where pieces of equipment (which may be bonded together or not) are connected to an earthing point. This ensures any

145

electrical charge in the equipment is given the means to constantly flow to earth, thus ensuring there is no potentially dangerous build-up of charge which could lead to a sudden discharge of electricity, by way of a spark.

Some other considerations are:

▶ Incidental objects and equipment, such as probes, thermometers and spray nozzles, which are isolated, but which can become sufficiently charged to cause a static spark, may need special consideration.
▶ The cables used for bonding and grounding cables should be heavy duty. This is to ensure that they can cope with physical wear and tear without compromising their grounding ability. It is also to ensure that their electrical resistance is as low as possible.
▶ The bonding of process equipment to conductors must be direct and positive.
▶ Using inert gas, such as nitrogen, within the ullage space of a storage vessel will prevent an explosion or flash fire occurring if an electrostatic spark should occur. The inert gas lowers the oxygen content of the gas in the ullage space, thus ensuring there is insufficient oxygen to support a burning process (oxygen being part of the fire triangle).
▶ Operators should wear anti-static clothing.

The identification of ignition sources

Potential ignition sources need to be considered when conducting a risk assessment. These include:

▶ Smoking and smoking material
  ▷ A total ban on smoking and the taking of smoking materials into controlled areas should be enforced.
▶ Vehicles
  ▷ Vehicles may be totally prohibited or restricted to only specially adapted vehicles.
▶ Hot work such as welding, grinding, burning, etc.
  ▷ Implement a permit-to-work regime.

- Electrical equipment
  - The equipment should be suitable for the zone it is intended to be used in. It should also be properly and regularly inspected and maintained.
- Machinery such as generators, compressors, etc.
- Hot surfaces such as those heated by process or by local weather (hot deserts)
- Heated process equipment such as dryers and furnaces
- Flames such as pilot lights
- Space heating equipment
- Sparks from lights and switches
  - Use only electrical equipment and instrumentation classified for the zone in which it is located.
- Impact sparks
- Stray current from electrical equipment
  - Ensure all equipment is bonded and earthed.
- Electrostatic discharge sparks
  - Bond and ground all plant and equipment.
- Electromagnetic radiation
  - The correct selection of equipment is important to avoid high intensity electromagnetic radiation sources, e.g. limitation on the power input to fibre optic systems, avoidance of high intensity lasers or sources of infrared radiation.
- Lightning
  - There should be measures in place which reduce the potential of a lightning strike as well as a grounding system to disperse any charge that may affect the installation. A further consideration is to look at weather windows (i.e. to not work during electrical storms).

Other control measures include:

- Controls over activities that create intermittent hazardous areas, e.g. tanker loading/unloading.
- Control of maintenance activities that may cause sparks or flames through a permit-to-work system.
- Precautions to control the risk from pyrophoric scale. This is where a substance can ignite spontaneously in air, particularly

147

humid air, and is usually associated with the formation of ferrous sulphide.

▶ Where control and/or detection equipment is regarded as critical, such as smoke and flame detectors, then a back-up or secondary system may be considered appropriate.

All of these control measures are supplementary to the main control and fire-fighting systems such as emergency shutdown systems, fire deluge systems, sprinkler systems, etc.

Zoning/hazardous area classification and selection of suitable ignition-protected electrical and mechanical equipment and critical control equipment

Zoning

A place where an explosive atmosphere may occur on a basis frequent enough to be regarded as requiring special precautions to reduce the risk of a fire or explosion to an acceptable level is called a 'hazardous place'.

A place where an explosive atmosphere is not expected to occur on a basis frequent enough to be regarded as requiring special precautions is called a 'non-hazardous place'.

Determining which areas are hazardous places, and to what extent, is called a 'hazardous area classification study'.

A hazardous area classification study involves giving due consideration to the following:

▶ The flammable materials that may be present
▶ The physical properties and characteristics of each of the flammable materials
▶ The source of potential releases and how they can form explosive atmospheres
▶ Prevailing operating temperatures and pressures
▶ Presence, degree and availability of ventilation (forced and natural)
▶ Dispersion of released vapours to below flammable limits
▶ The probability of each release scenario

Consideration of these factors will enable the appropriate selection of zone classification for each area regarded as hazardous, as well as the geographical extent of each zone.

Hazardous areas are classified into zones based on an assessment of two factors:

1 The frequency of occurrence of an explosive gas atmosphere
2 The duration of an explosive gas atmosphere

These two factors in combination will then facilitate the decision-making process to determine which zone will apply to the area under consideration.

▷ Zone 0: An area in which an explosive gas atmosphere is present continuously or for long periods of time
▷ Zone 1: An area in which an explosive gas atmosphere is likely to occur in normal operation
▷ Zone 2: An area in which an explosive gas atmosphere is not likely to occur in normal operation but, if it does occur, will only exist for a short period of time

Selection of equipment

Apparatus, tools and equipment are categorized in accordance with their ability to meet the standards required when used within each zone according to Table 15.1 below.

As well as taking into account the sparks that electrical equipment can generate, consideration also needs to be given to the potential surface temperature of all equipment, not just electrical equipment, although most electrical equipment does generate heat as a matter of course.

In order to facilitate this, temperatures have been categorized into six classes, T1–T6. The bigger the T-number, the lower the allowable temperature of any equipment used. The temperature class will be determined by the auto ignition temperature of the substance involved.

149

Table 15.1 – Tools and equipment categorization in zoned areas

Zone 0	Zone 1	Zone 2
An area in which an explosive gas atmosphere is present continuously or for long periods of time	An area in which an explosive gas atmosphere is likely to occur in normal operation	An area in which an explosive gas atmosphere is not likely to occur in normal operation but, if it does occur, will only exist for a short period of time
Category 1 equipment	Category 2 equipment	Category 3 equipment
Note: Although this equipment is categorized for use in Zone 0, it can also be used in Zones 1 and 2	Note: Although this equipment is categorized for use in Zone 1, it can also be used in Zone 2	Note: This equipment can only be used in Zone 2
'ia' – Intrinsically safe	'd' – Flameproof enclosure	Electrical type 'n'
Ex s – Special protection if specifically certified for Zone 0	'p' – Pressurized	
	'q' – Powder filled	
	'o' – Oil immersion	
	'e' – Intrinsically safe	
	'ib' – Intrinsically safe	
	'm' – Encapsulated	
	's' – Special protection	

Source: Adapted from 'Hazardous Area Classification and Control of Ignition Sources' available at www.hse.gov.uk/comah/sragtech/techmeasareaclas.htm

Table 15.2 – Temperature classification for tools and equipment in zoned areas

Temperature classification	Maximum surface temperature	Substances can be used which will not auto ignite at temperatures below
T1	450°C	450°C
T2	300°C	300°C
T3	200°C	200°C
T4	135°C	135°C
T5	100°C	100°C
T6	85°C	85°C

Source: Adapted from 'Hazardous Area Classification and Control of Ignition Sources' available at www.hse.gov.uk/comah/sragtech/techmeasareaclas.htm

Sub-element 3.5: Fire hazards, risks and controls

Learning outcome

Outline the fire hazards, risks and controls relating to hydrocarbons.

 Revision exercise

Write your answers to the questions below on a separate sheet of paper without referring to the information in this book in the first instance. Once you have answered all of the questions, you can refer back to the revision guide to compare your answers and this will give you an indication of how much knowledge you have been able to absorb or whether you need to revise this section further.

Q1 **Give TWO** examples of potential threats from a lightning strike.

Q2 **Give** the **THREE** elements which make up the fire triangle.

Q3 **Explain** what 'thermal radiation' is.

Q4 **Describe** how electrostatic charges can be generated within hydrocarbon liquid product.

Q5 **Outline THREE** factors which can influence the build-up of static charge within hydrocarbon liquid product.

Q6 **Describe ONE** control measure that can be implemented to reduce static charge in piping systems.

Q7 **Describe TWO** control measures that can be implemented to reduce static charge when filling operations take place.

Q8 **Explain** why filtration of hydrocarbon liquid product can create high levels of electrostatic charge.

Q9 **Explain** how bonding and grounding of equipment can help to reduce the potential for static charges to be generated.

Q10 **Give FIVE** examples of sources of ignition.

Q11 **Explain** what 'zoning' of hazardous areas is and why it is implemented.

Q12 **Explain** why electrical tools need to be categorized as suitable for use in specifically zoned hazardous areas.

Sub-element 3.6:
Furnace and
boiler operations

Learning outcome

Outline the hazards, risks and controls available for operating boilers and furnaces.

🔑 Key revision points

Use of furnace and boiler operations ☐

Hazards and risks of operating boilers and furnaces, in particular those arising from loss of pilot gas supply, over-filling, flame impingement, firebox overpressure, low tube flow, control of tube metal temperature (TMT) ☐

Use of furnace and boiler operations

Introduction

Boilers are devices which heat large quantities of water in order to provide a constant supply of hot water, or to turn it into steam. Where steam is generated, this is captured and kept in a pressurized state.

Furnaces, or process heaters are devices which are used to provide a large source of heat to various process streams and are used extensively in the oil and gas industry.

Boiler/furnace components

▷ The combustion chamber is where the heat is generated by burning the fuel.
▷ The heat exchanger is where the heat is transferred to water (or product if it is a furnace/process heater).
▷ The chimney or flue allows exhaust gases to escape to atmosphere. Some boilers/furnaces also have a heating coil positioned within the flue. This extracts some of the heat from the exhaust gases which would otherwise be lost to the atmosphere. These devices are known as waste heat heaters or economizers.
▷ The controls, of which there are a number, allow the heating of water or product to be done in a regulated, efficient and safe manner. The combustion and operating controls regulate the rate of fuel used to meet the demand.

Boiler/furnace safety components

All boilers used for steam production have a safety relief valve which allows excessive steam pressure to be released to prevent overpressure or explosion.

Boilers also have a drain which allows sediments and contaminates to be drained from the water.

Other boiler/furnace safety controls include automatic and continuous monitoring of pressure and temperature, high and low gas or oil pressure, high and low water/product levels, and flame

154

safeguard controls. Generally, these are connected to a circuit breaker which prevents the firing of the boiler if any of these safety controls are activated.

Boiler types

There are two main types of boiler: fire tube boilers and water tube boilers. As their names indicate, one has fire within its tubes, whereas the other has water within its tubes.

Boiler types – fire tube boilers

A fire tube boiler is one which generates hot gases which then pass through a number of tubes before being expelled out of the flue. These tubes run through a sealed and insulated container of water and the heat from the gases is transferred by thermal conductivity to the water which then turns to steam. The steam from the boiler then exits through a tube at the top of the container.

Boiler types – water tube boilers

Water tube boilers have basically the opposite configuration of fire tubes. In a water tube boiler, a number of tubes run through the furnace part of the boiler. This heats the water inside the tubes which turns it to steam.

Furnace types

There are two main types of furnaces/process heaters: natural draught and balanced draught.

Furnace types – natural draught

This type of furnace/process heater uses the principle of a natural draught to move the air and combustion gases through the combustion chamber. It has a high chimney stack which creates a difference in pressure between the cold air at the burner and the hot and less dense air at the top of the chimney stack. This creates a natural flow of combusted air up the chimney.

This type of furnace/process heater has a fan or blower to increase and control the flow of air and combustion gases through the combustion chamber.

Hazards and risks of operating boilers and furnaces in particular those arising from loss of pilot gas supply, over-filling, flame impingement, firebox overpressure, low tube flow, control of tube metal temperature (TMT)

Boiler/furnace hazards

There are a number of specific hazards associated with operating boilers and furnaces, and these need special attention.

Boiler/furnace hazards – pilot lights

Pilot lights provide a source of ignition to the main boiler when it needs to be fired up. In a situation where the pilot light fails to light or goes out, gas will continue to enter the chamber, causing a build-up of flammable gas which if ignited could cause an explosion.

In order to counteract this scenario, a sensory device called a 'thermocouple' is located in close proximity to the pilot light in order to detect if there is heat coming from the pilot light. If no heat is detected, the device will activate a relay which will close the main gas valve.

Boiler/furnace hazards – over-firing

Over-firing the boiler is basically allowing the heat flux to increase to a level beyond its upper Maximum Continuous Rating (MCR), which is set by the manufacturer of the boiler. This can then have an impact on, amongst other things, the furnace walls and the surface temperature of the refractory. It can also result in a substantial increase in tube and membrane operating temperatures, which can lead to a degradation of tube metallurgy and strength.

Boiler/furnace hazards – flame impingement

This is where the flame produced by the burners comes into contact with the surface being heated. If this occurs there tends to be a gradual build-up of carbon on the inside of the tube at the point where the flame is in contact. If this process is allowed to continue, it can lead to the tube eventually becoming blocked, resulting in the potential rupture of the tube.

Causes of flame impingement can be due to:

▶ Improper burner adjustment
▶ Poor forced draught conditions
▶ Poor design

Control measures

The first step is to ensure the flame is kept off the heated surface. Regular inspections in order to monitor flame behaviour are essential.

Boiler/furnace hazards – firebox overpressure

This typically occurs after a flameout, which is where the burner flame is extinguished for some reason. The fumes, gases or vapour from the fuel will begin to build up inside the combustion chamber, which will invariably be hot. This will make them highly volatile, and when they reach their Lower Explosive Limit (LEL) and make contact with a source of ignition, this will cause overpressure (explosion).

Boiler/furnace hazards – low tube flow

Normal circulation within water or steam pipes heated by a boiler is generated by the difference in density between cooled water and hot water/steam. Any adverse conditions within the flow system will affect this flow rate and create a situation where flow rates are eventually reduced. When this becomes acute, it is known as 'low tube flow'.

Control measures include:

▶ Flushing the tubes regularly to ensure any blockages are removed
▶ With regard to boilers, keeping the water quality at the recommended level

157

▷ Ensuring flame impingement does not occur
▷ With regard to furnaces/process heaters, using tube pass flow
meters to monitor product flow

Boiler/furnace hazards – Tube Metal Temperature (TMT)

This is where localized overheating occurs which can lead to potential component failure. It is generally due to poor water quality, where suspended material tends to congregate in the bottom of the boiler and cause scale to develop. It can also be caused by ingress of product (e.g. oil) into the condensate return system.

Controls include regular testing of water and blowing down on a regular basis, as well as checking for leaks in the water/steam circulation system.

Maintenance

Maintenance – boiler blow down

Total Dissolved Solids (TDS) are substances such as minerals, salts and even metals, which are held in a suspended form within water. If these solids are of a sufficient concentration within the water used in a boiler system, they can attach themselves to the inside of boilers and, over time, build up to form scale.

Consequently, the first action is to maintain the solids below a certain limit. This is done by testing the water with a Total Dissolved Solids (TDS) meter or conductivity meter. This measures the conductivity of the water, which is an indication of the measure of TDS within the water. As the TDS concentration increases, the likelihood that the dissolved solids will precipitate out of the water and form scale also increases. At this point it is necessary to drain some of the water from the system, called boiler blow down, in order to remove some of those dissolved solids and keep the TDS concentration below the level where they will precipitate.

Maintenance is a critical factor in maintaining a safe and efficient boiler or furnace. A log should be kept of all inspections and maintenance activities, and inspections should be carried out against a checklist to ensure the equipment is operating properly.

A daily checklist might comprise the following:

▶ Boiler/furnace use and sequencing
▶ Overall visual inspection
▶ Lubricate all oil and grease nipples
▶ For boilers, check steam pressure
▶ For boilers, check water level
▶ Check burner
▶ Check motor
▶ Check air temperature
▶ Check oil filter
▶ Inspect oil heaters
▶ For boilers, check water treatment
▶ Complete log

There will be other checklists for activities which need conducting on a weekly, monthly and annual basis.

Sub-element 3.6: Furnace and boiler operations

Learning outcome

Outline the hazards, risks and controls available for operating boilers and furnaces.

✎ Revision exercise

Write your answers to the questions below on a separate sheet of paper without referring to the information in this book in the first instance. Once you have answered all of the questions, you can refer back to the revision guide to compare your answers and this will give you an indication of how much knowledge you have been able to absorb or whether you need to revise this section further.

Q1 There are two main types of boiler. **Describe** how both work and how they differ.

Q2 **Outline** the hazards associated with pilot lights in boilers and furnaces and the safety measures which will address these hazards.

Q3 **Explain** what 'over-firing' is with regard to boilers and furnaces.

Q4 **Explain** what 'flame impingement' is with regard to boilers and furnaces, and how it can be controlled.

Q5 **Explain** what 'firebox overpressure' is with regard to boilers and furnaces.

Q6 **Explain** what 'low tube flow' is with regard to boilers and furnaces, and how it can be controlled.

Q7 **Explain** what 'tube metal temperature' is with regard to boilers and furnaces, and how it can be controlled.

Q8 **Explain** what 'Total Dissolved Solids (TDS)' are with regard to boilers and furnaces, and how they can be controlled.

Element 4

CHAPTERS 17–18

Sub-element 4.1: Fire and explosion in the oil and gas industries

Learning outcome

Outline appropriate control measures to minimize the effects of fire and explosion in the oil and gas industries.

🔑 Key revision points

Leak and fire detection systems ☐

Passive fire protection ☐

Active fire protection systems – water based extinguishing systems ☐

Active fire protection systems – chemical/foam based extinguishing systems ☐

Active fire protection systems – inert based extinguishing systems ☐

Examples of fire protection systems and their function for equipment specific types ☐

Leak and fire detection systems

Detection systems

The Fire and Gas detection system (F&G) is a combination of various devices strategically positioned throughout the installation which aim to give the earliest possible warning of any potential problem that could escalate into a fire or explosion if it is not dealt with as quickly as possible.

Detection systems – spot fire detection systems

Spot fire detection systems are designed to conduct a continuous thermal surveillance of a specific area. They are sensitive to infrared radiation within a cone of vision.

Detection systems – camera based flame detectors

Camera based flame detectors are capable of detecting and pinpointing a fire from long distances as soon as the fire starts. These pictures can be relayed back to the control room operator.

Detection systems – ultraviolet flame detectors

Ultraviolet flame detectors rely on detecting the ultraviolet radiation produced by flames. However, the more advanced ultraviolet flame detectors are capable of detecting hydrogen fires.

Detection systems – combined ultraviolet and infrared flame detectors

These detectors are capable of detecting both ultraviolet radiation and infrared radiation. They tend to be used where the reduction of false alarms is an issue.

Detection systems – point heat detectors

These systems are capable of detecting a rise in temperature at any point within the range of the detector.

Detection systems – linear heat detectors

Linear heat detectors are used where heat detection is required in a linear fashion. It comprises of a cable which is made up of a pair of twisted wires which are each sheathed in a polymer coating. This coating is engineered to break down at a specific temperature, allowing the twisted wires to make contact and send a signal to the alarm box and/or control room.

Detection systems – infrared absorption combustible gas detectors

These work by being able to recognize the specific absorption characteristics that hydrocarbon molecules have to infrared light. The more hydrocarbon molecules that are present in a given space being monitored by these devices, the higher the absorption of infrared radiation.

Detection systems – point infrared gas detectors

These monitor a specific location with the aim of measuring the concentration of gas at that location. Air is drawn from the sampling location and passed over the detection chamber within the device.

Detection systems – open path infrared gas

These work on the principle of sending two infrared beams between the transmitter and the receiver. The first, a sample beam, is the infrared beam which is set to detect the presence of hydrocarbons. The second infrared beam is set so that its wavelength is outside the gas absorbing range. The two beams are constantly compared and whilst the comparison remains the same as the factory setting, no gas is present and no alarm is raised. However, if the comparison changes because some of the sample beam has been absorbed by gas, the device will trigger an alarm.

Detection systems – ionization point smoke detectors

These use a small radioactive source (a radioisotope) to ionize air inside a detection chamber. Any change in this ionization, due to the presence of smoke particles, will activate the alarm.

Detection systems – optical point smoke detectors

These are designed to detect smoke particles inside a sampling chamber. This is where a source of light is collimated into a beam and a photoelectric sensor is set at an angle to the beam to act as a light detector and monitor it.

Detection systems – optical beam smoke detectors

These work on the principle of measuring any difference in light emitted from one point and received at another point. Where smoke particles absorb or scatter the light, this is detected and the alarm is triggered.

Detection systems – aspiration smoke detectors

These use a network of tubes to draw samples of air from various locations to a central detection unit. This detection unit then looks for the presence of smoke particles by measuring if light passed through the samples is scattered or not. If it is scattered, this indicates the presence of smoke and the alarm will be triggered.

Detection systems – leak detection systems

These work on the principle of detecting the ultrasonic sound emitted when a leak of gas or vapour occurs from a small gap, such as those that may develop on valves, flanges and joints in pipework. This sound is inaudible to the human ear but is highly characteristic when gas or vapour leaks from the gap in the plant.

Other design considerations for alarm systems

Buildings, plant and equipment layout tend to be divided into three categories as far as fire and leak safety engineering is concerned: fire compartments, detection zones and alarm zones.

Design considerations – fire compartments

Buildings are generally divided into sections, known as fire compartments, which use fire resistant structures (walls and floors) to enclose them in order to limit the spread of any fire that may break out within any one of these sections.

Design considerations – detection zones

These are essentially a convenient way of dividing up a building, area or plant into manageable sections so as to assist in quickly locating the position of any fire or leak. The size and position of the detection zones will be dependent on such things as the extent of areas used for one particular process or the number of people working in one section at a time.

Complex buildings, where it is necessary to operate alarm devices differently in various parts of the building, should be divided into alarm zones so that all of the alarm devices in one alarm zone operate in the same way. If the only requirement is to activate all the alarms in order to provide a single common evacuate signal once a fire is detected, then alarm zones are not needed. In this case the whole building is regarded as one alarm zone.

Passive fire protection

Introduction

Passive fire protection covers the materials, products and design measures built into a building or structure in order that any fire which may start in the building or structure is restricted in its growth and spread. This is achieved by controlling the flammability of the structure, including walls, ceilings, floors, doors, etc., as well as by protecting structural steel members from severe heat which might compromise their integrity. Finally, the building or structure is designed in a way that divides it into separate fire containing compartments. This is so that any fire which may start will be restricted, or contained within a limited area.

Types of passive fire protection

There are many types of Passive Fire Protection (PFP) materials available. These can be broadly categorized as follows:

- Spray-applied coatings
- Blanket/flexible jacket/wrap around systems
- Prefabricated sections such as walls
- Enclosures and casings
- Composites
- Seals and sealants
- Systems (e.g. cable transit blocks, inspection hatches, pipe penetration systems through bulkheads)

Aims of passive fire protection

Passive fire protection has three main aims:

1. To prevent steel structures, which are load-bearing, reaching a temperature where their integrity will be compromised.
2. To protect process vessels and their supportive structures (legs).
3. To prevent heat transfer through walls, floors and ceilings into adjacent rooms or spaces by limiting the inner wall temperature to no more than 180°C in any one location.

Types of fire

Fires can be categorized as being either cellulosic or hydrocarbon. This refers to the fuel which is being consumed by the fire. Cellulosic fires are those which burn materials such as timbers, upholstery or paper. Hydrocarbon fires are those which burn oils and fuels.

Coding of fire walls

Walls and divisions which have been manufactured as passive fire protection are coded in relation to the type of fire they are designed to withstand and its duration. The code is made up of a letter (either H or A) followed by a number (0, 30, 60 or 120).

The number which follows the letter indicates the length of time the wall or division has been designed to hold back the fire. For example, a wall with a 'H60' coding is designed to withstand a hydrocarbon fire for 60 minutes, whilst a wall with an 'A30' coding is designed to withstand a cellulosic fire for 30 minutes.

Active fire protection systems – water based extinguishing systems

Active fire protection systems are those fire protection systems which are primed and ready to be activated on a given signal.

Water deluge system

This is a means of fighting a fire with large amounts of water. It works by means of a series of nozzles connected to a piping network. The nozzles are kept permanently in the OPEN position. When an emergency situation arises, the deluge system is activated by the emergency shutdown system. This switches on the dedicated fire water pumps, resulting in water being emitted from the nozzles and deluging the area.

Sprinkler system

This consists of a grid of water pipes, usually located within a roof void, which have sprinkler heads located at regular intervals that protrude through the roof. The sprinklers are spaced in such a way as to give adequate cover to the whole area which needs protecting against the outbreak of fire.

The water in the pipe system remains under pressure (active) but is held back at each sprinkler head by a thermal element. When a thermal element is activated because it has reached its predetermined temperature, it allows water to be sprayed from that sprinkler head. Only sprinkler heads in the vicinity of the fire will be activated, thus ensuring only the immediate area around the fire is sprayed with water. This minimizes any possible water damage.

Active fire protection systems – chemical/foam based extinguishing systems

Foam based extinguishing systems

Foam is used as a means to extinguish liquid fires. Foam is made by combining water with a foam concentrate and mixing it in a way that introduces air. It is basically a stable mass of small air-filled bubbles with a density which is lower than oil or gasoline. This means it will readily flow over the surface of burning fuel.

Controlling effects of foam

Controlling effects of foam – a separating effect

The foam cover separates the combustion zone from the surrounding atmosphere and, as such, prevents oxygen feeding the fire.

Controlling effects of foam – a covering effect

The foam cover stops the evaporation of flammable vapours from the burning material. This means flammable vapours, which would normally be encouraged by the heat of the fire and which feed the fire, are stopped at source.

Controlling effects of foam – a cooling effect

When foam is applied to a liquid fire, water is gradually discharged by the foam, which adds a cooling effect to the flame.

Controlling effects of foam – a suppression effect

High- or medium-expansion foam used to flood areas will prevent the release of flammable vapours/gases from combining with air which is necessary for the combustion process to take place.

Controlling effects of foam – an insulating effect

Foam has a low thermal conductivity level. Consequently, when foam covers any flammable liquid which is not burning, it tends to insulate it from thermal radiation and ignition.

Controlling effects of foam – a film formation effect (AFFF – Aqueous Film Forming Foam)

Foam which is used on non-polar hydrocarbons (the majority of hydrocarbons) produces a thin aqueous film which helps the foam flow, as well as assisting in the extinguishing process and inhibiting re-ignition.

171

Polymer film additives

A small group of hydrocarbons, known as polar hydrocarbons, will destroy any ordinary foam used as an extinguishing agent as soon as it is applied. In order to combat this reaction, an additive called a polymer film former is added to the foam. When the foam with the additive is applied to the burning hydrocarbon, the film floats on top of the hydrocarbon acting as a barrier between the hydrocarbon and the foam, thus stopping the hydrocarbon from breaking down the foam.

Chemical fire extinguishing systems

These use dry powder to extinguish a fire. They can be deployed as hand held extinguishers for use on localized small fires or fixed systems to cover a specific area. Both the hand held and fixed systems use the principle of introducing an inert gas, usually nitrogen, into the base of the vessel holding the dry chemical. This pressurizes the vessel and allows the dry chemical particles to exit through a discharge valve at a controlled velocity and in a form where they are suspended in the inert gas (a cloud). The chemical reacts with the fire and extinguishes it. Monoammonium phosphate based chemicals melt at a relatively low temperature, thus blanketing the burning surfaces and preventing re-ignition.

Fixed dry chemical powder fire extinguishing systems on bulk gas carrying ships

In 2009 the International Maritime Organization approved guidelines for installing fixed dry chemical powder fire extinguishing systems on board ships carrying liquefied gases in bulk. The aim is to protect on-deck cargo areas of ships carrying liquefied gases in bulk in accordance with Safety of Life at Sea (SOLAS) regulation II-2/1.6.2 and chapter 11 of the International Code for the Construction and Equipment of Ships Carrying Liquefied Gases in Bulk (IGC Code).

172

Active fire protection systems – inert based extinguishing systems

Inert gas systems

These are based on the principle that fire needs oxygen in the air to continue its combustion process (burn). The level of oxygen in the air is normally 21 per cent, but if this falls below 15 per cent, the combustion process will stop as there is not enough oxygen present to feed the fire. Inert gas systems aim to reduce the oxygen in the target area to below 15 per cent, thus extinguishing the fire.

The gases used in these systems are carbon dioxide (CO_2), nitrogen (N) and argon (Ar), or a combination of these gases.

Water mist systems

These work on the same principle as a sprinkler system. However, when activated, the water that is discharged is in micron sized droplets which cover a wider area. The droplets rapidly convert the energy in the fire to steam, which starves the fire of oxygen. Because the droplets are so small and abundant, the water absorbs the energy of the fire much faster than a sprinkler system and uses less water. This in turn reduces the potential for any damage that may be caused to the surrounding area.

Examples of fire protection systems and their function for equipment-specific types

Fire protection systems for floating roof tanks

Storage tanks with floating roofs have the potential to leak vapour from the rim seal. If that happens and the vapour ignites and the situation is not dealt with immediately, it can result in an escalation of events and a potentially catastrophic outcome.

One solution is to place a linear heat detection system at the rim seal area of the roof and connect this to a number of foam based fire extinguishing modules positioned around the perimeter of the roof. When the linear heat detector registers a fire, as well as raising the alarm, it will also activate the fire extinguisher module nearest to the fire. This will then flood the immediate area with foam and extinguish the fire.

Fire protection systems for spherical storage tanks

Fire protection systems for spherical storage tanks generally employ a water based fire protection system.

Fire protection systems for spherical storage tanks – water deluge system

A water deluge system for spherical storage tanks works on the principle of having a single or a number of outlets placed on the top of the vessel. When these are activated by an alarm which has sensed a fire, water pours from these outlets and runs down over the surface of the vessel, potentially extinguishing the fire.

Fire protection systems for spherical storage tanks – fixed monitor system

A fixed monitor system for spherical storage tanks is where vessels are surrounded by ground based water hydrants. These either have monitors affixed to them, or monitors are positioned independently around the vessel so that a visual impression is available to the control room at all times.

When a fire is detected, the alarm is raised and the hydrants are either automatically operated by the system or manually operated by personnel. The hydrants can be fitted with an oscillating facility which sprays the water from side to side. This allows the water to be directed over a wide area without the need for an operator to be present.

Fire protection systems for spherical storage tanks –
water spray system

A water spray system for spherical storage tanks is made up of a
network of water pipes which totally surround the vessel. These
have spray nozzles fitted to them at regular intervals which point
inwards towards the vessel. When these are activated by an alarm
which has sensed a fire, water is sprayed from these nozzles onto
the entire surface of the vessel with the aim of extinguishing the fire.

Fire protection systems for gas turbine and compressor systems

A complete gas turbine system is made up of a number of separate
areas or compartments. These include the turbine compartment, the
generator compartment, the fuel pump compartment, the lubricating
skid and the electrical control room, all of which will require individual
fire protection systems.

However, the major fire hazard is associated with the turbine
compartment, where there is the potential for fuel or lubricating oil
to leak and come into contact with surfaces of the turbine which
operate at temperatures well above the auto-ignition temperature of
the fuel and oil. If either the fuel or the lubricating oil leaks and comes
into contact with one of these hot surfaces, it will cause a fire.

The most appropriate means of dealing with a fire in the turbine
compartment is with a water mist system. The mist nozzles are
positioned in a location where they will not directly spray the surface
but create a mist cloud which envelopes and cools these hot
surfaces by heat transfer, turning the water droplets to steam.

Sub-element 4.1: Fire and explosion in the oil and gas industries

Learning outcome

Outline appropriate control measures to minimize the effects of fire
and explosion in the oil and gas industries.

175

✎ Revision exercise

Write your answers to the questions below on a separate sheet of paper without referring to the information in this book in the first instance. Once you have answered all of the questions, you can refer back to the revision guide to compare your answers and this will give you an indication of how much knowledge you have been able to absorb or whether you need to revise this section further.

Q1 **Give FIVE** examples of components within a fire and gas detection system.

Q2 **Explain** the following:
   (a) fire compartments
   (b) detection zones
   (c) alarm zones

Q3 **Explain** what is meant by the phrase 'passive fire protection'.

Q4 Fire walls are coded to indicate their ability to withstand fire. **Explain** how this code is made up and what it represents.

Q5 **Explain** what is meant by the phrase 'active fire protection system'.

Q6 Foam, when used as a means for fire-fighting, has a number of controlling effects. **Outline THREE** effects.

Q7 **Describe** how a chemical fire extinguishing system using dry powder works.

Q8 **Explain** how a water mist fire extinguishing system works.

Q9 **Give** an example of a fire protection system for storage tanks.

Sub-element 4.2:
Emergency
response

Learning outcome

Outline the principles, procedures and resources for effective
emergency response.

🔑 **Key revision points**

Emergency plan ☐

Alarms – importance of response ☐

Medical emergency planning, tiered response, medical
evacuation procedures and back up resources ☐

Principles of escape, evacuation and rescue from onshore
facilities and offshore platforms ☐

Emergency plan

On-site emergency plan

Operators of both onshore and offshore installations should undertake the following actions as part of the procedure of writing a comprehensive emergency plan:

▷ Identify all the major hazards associated with the operations together with their source, type, scale and consequences. This should include malicious acts.
▷ From these hazards, identify all the conceivable scenarios that could arise which will need an emergency response, including those which involve evacuation, escape and rescue.
▷ From these scenarios, produce a well-defined plan of action which establishes the appropriate response to an emergency and which takes into account the varying demands of different scenarios.
▷ Establish the procedures and frequencies required to test and practise the emergency response to be followed in each of the different scenarios identified.
▷ Establish the formal command structure.
▷ Establish what those people who will be expected to take an active part in any emergency response, including those in overall command, are competent to do.

178

▷ Establish that there are enough people to respond to any emergency.
▷ Establish the roles and responsibilities of all individuals on the installation.
▷ Establish that there are enough resources to respond to any emergency.
▷ For onshore sites, establish that there are plans for both on site and off site should the emergency not be contained within the site.
▷ For onshore sites, establish that a Major Accident Prevention Policy (MAPP) has been prepared and is current.
▷ Establish what measures will be necessary to facilitate a site clean-up and remediation following an incident.

Contents of an emergency plan

Responsibilities and authority of those overseeing an emergency

The command structure for managing the on-site response to an emergency in accordance with the planned scheme. This will include:

▷ The names and positions of persons authorized to set emergency procedures in motion.
▷ The name and position of the person in charge of, and co-ordinating, the on-site mitigatory action.
▷ The name and position of the person responsible for liaising with external agencies and/or local authorities.
▷ Details of what arrangements have been made for occasions when senior managers are not available.
▷ The contact details of all authorized personnel.

Types of events planned for and extent of responses planned

This is the principle aspect of an emergency plan. It should include details of the following issues:

▷ The types of emergency which have been regarded as reasonably conceivable.

179

▶ The response strategy for each of these situations.
▶ The details and responsibilities of personnel who have been allocated roles to play in an emergency.
▶ The details and location of any special equipment, such as fire-fighting equipment and damage control facilities.

Alarm systems and responses to alarms

The plan should include what alarm systems and arrangements have been made for early detection of a potential emergency. This will include what arrangements have been made and responsibilities for ensuring an appropriate response is made by personnel, such as evacuating the area or facility, taking shelter, using protective equipment, etc.

Arrangements for triggering any off-site emergency plan

Where an off-site plan is applicable, it must include details of what arrangements have been made for alerting off-site emergency services and other agencies such as water companies, environmental agencies, etc. Also, under what circumstances these alerts should be made, what information will be required by each service or agency, and their contact details, should all be planned for.

Training and instructions

This section of the emergency plan covers the arrangements made for training staff in their roles and responsibilities in an emergency. It also covers the arrangements for, and frequency of, conducting exercises based on all the identified emergency scenarios.

Finally, the plan should include details of how contractors and visitors will be given instructions of how they should respond in an emergency.

Off-site communication measures

It should be established who will be responsible for contacting and briefing the media, and media contact details should also be included.

Fire and explosion strategy

This is a combination of measures taken to reduce the risk to personnel in the case of fire or explosion, or that reduce the risk of fire and explosion happening in the first place. Some of those measures listed here will apply specifically to offshore installations, others will apply to onshore installations, and the rest to either type of installation.

Measures to be considered in formulating a fire and explosion strategy for a specific installation include:

▷ Buildings which are occupied should have an assessment made of the risks and hazards they might be vulnerable to if a major incident occurred. From that assessment, appropriate measures should be taken to address those issues. The buildings included in the assessment process should also include temporary and secondary refuges.

▷ Escape routes should be clearly marked using high visibility signage along their entire route.

▷ Escape routes should be well lit and include a contingency for emergency lighting in case of power loss.

▷ All escape routes,where appropriate, should be protected by fire walls or by deluge fire protection systems.

▷ Escape routes should be of a size that is adequate to accommodate all personnel.

▷ Where appropriate, the installation should be compartmentalized (have fire walls between compartments).

▷ Where appropriate, blowout, or explosion panels, should be strategically positioned within the installation to alleviate any overpressure.

▷ Where appropriate, escape routes should have heat activated deluge/sprinklers within them.

▷ Each area of the installation should have more than one escape route.

▷ Escape routes should be protected against the effects of fire and explosion.

▷ There should be internal access to the helideck from any temporary refuge facility.

181

▷ There should be a policy of ensuring the number of overrides and inhibits applied to the Emergency Shutdown (ESD) system and the Fire and Gas (F&G) system is kept to a minimum.

▷ At the design stage of an emergency shutdown system, failsafe and fireproof ball valves should be incorporated so their integrity will not be compromised in the case of a fire.

▷ At the design stage of a process system, the amount of flanged pipework should be minimized in order to reduce the potential for leaks.

▷ At the design stage the inventory of hazardous substances should be reduced to a minimum.

▷ Emergency Shutdown Valves (ESDVs) should be enclosed with fireproof casing.

▷ Water deluge operating skids should be situated away from the area they are protecting.

▷ Access doors to accommodation areas should have automatic door closers fitted to prevent ingress of smoke and flames.

▷ All enclosures which house rotating equipment and electric drives should have Very Early Smoke Detection Apparatus (VESDA) fitted.

▷ Measures should be taken to ensure that mechanical and natural ventilation to production areas is sufficient to assist in dispersing any gas leak.

▷ The accommodation and control rooms should be segregated and distanced away from production processes wherever possible.

▷ The control rooms and emergency command and control centres should be segregated using blast and fire walls.

▷ Subsea Isolation Valves (SSIVs) should be fitted in sea lines and wells.

▷ High Integrity Pipeline Protection Systems (HIPPSs) should be fitted where appropriate.

▷ External fire protection should be fitted to the accommodation rooms and Temporary Refuge (TR).

▷ The Temporary Refuge (TR) should be airtight and always under positive atmospheric pressure.

▷ There should be a separate Emergency Command and Control (ECC) centre in the Temporary Refuge (TR) when the control room is not situated within the TR.

Alarms – importance of response

Onshore alarms

When an emergency arises within an onshore facility, it is imperative that everybody on the site, as well as the general public in the vicinity, is made aware of the situation. This alarm signal is conveyed by means of a warning siren which is loud enough for everybody on the site, and in the surrounding area, to hear.

The alarm is also likely to be linked to the control centre of the emergency services so that they are automatically made aware of any emergency.

Offshore alarms

Offshore, there are two types of alarm used. The first type of alarm is the general platform alarm, which is a general alarm calling all personnel on board to go to their allocated muster station. The general platform alarm is an intermittent signal of a constant frequency.

The second type of alarm is the prepare to abandon platform alarm, which is sounded to inform personnel of the imminent evacuation of the platform. The prepare to abandon platform alarm is a continuous signal of variable frequency.

Both types of alarm are generally backed up with a public address announcement and by a visual alarm system where necessary.

Medical emergency planning, tiered response, medical evacuation procedures and back up resources

Medical response – onshore

Where a medical emergency occurs, the control and command team will be able to call upon the local ambulance service to deal with the casualties. Once they arrive on scene they will take control of managing the casualties by triaging them in order to prioritize their medical needs. Once their immediate medical needs have

183

been dealt with by the paramedics, they will be conveyed to hospital as required for more specific medical attention. Prior to the ambulance service arriving, casualties will be looked after by the first-aid team.

Medical response – offshore

The situation is different offshore because the support from external emergency services has much greater time constraints associated with it. Consequently, in the first instance, casualties will be dealt with by the platform medic, who will be trained in conducting triage of casualties. Triage is a means of prioritizing casualties in relation to their medical needs so that the most severe casualties can be dealt with first. The platform medic will also have the use of a fully equipped medical suite and be assisted by a number of first-aiders.

Depending on the location of the installation and the pre-arrangements for dealing with medical emergencies, the platform medic may also be able to call upon the service of an onshore doctor for advice.

As well as the platform medic, there is usually a person on the platform's standby vessel with Advanced Medical Aid (AMA) training who can be called upon if needed.

Search and Rescue (SAR) helicopters can also be called upon to medivac (medical evacuation) casualties from the installation. Search and rescue helicopters carry a paramedic who will attend to the immediate medical needs of the casualties. They will also assist in triaging the casualties where needed.

Once the immediate medical needs have been dealt with by the paramedics, those casualties needing further medical attention will be conveyed to hospital by the SAR helicopter.

Triage

This is a process of prioritizing casualties according to their medical needs.

Principles of escape, evacuation and rescue from onshore facilities and offshore platforms

Escape and evacuation – onshore

When personnel need to escape or evacuate an onshore installation there are a number of factors which can enhance their ability to escape without undue difficulty. These include:

▸ Escape routes should be clearly marked using high visibility signage along their entire route.

▸ Escape routes should be well lit and include emergency lighting in case power is lost.

▸ All escape routes should be protected, where possible, by fire walls or by deluge fire protection systems.

▸ Escape routes should be congestion-free (have clear access and egress) and be adequate in size to accommodate all personnel.

▸ Escape routes should have heat activated deluge/sprinklers within them.

▸ Each area of the installation should have more than one escape route.

▸ Where appropriate, the installation should be compartmentalized (have fire walls between compartments).

▸ Where appropriate, blowout, or explosion panels, should be strategically positioned within the installation to alleviate any overpressure.

Escape and evacuation – offshore

When personnel need to escape or evacuate an offshore installation, there are a number of factors which can enhance their ability to escape without undue difficulty. These include:

▸ Escape routes should be clearly marked using high visibility signage along their entire route.

▸ Escape routes should be well lit and include emergency lighting in case power is cut off.

185

▷ All escape routes should be protected, where possible, by fire walls or by deluge fire protection systems.

▷ Escape routes should be congestion-free (have clear access and egress) and be adequate in size to accommodate all personnel.

▷ The installation should be compartmentalized (have fire walls between compartments).

▷ Blowout, or explosion panels, should be strategically positioned within the installation to alleviate any overpressure.

▷ Escape routes should have heat activated deluge/sprinklers within them.

▷ Each area of the installation should have more than one escape route.

▷ Temporary Refuge (TR) buildings should be constructed from material which has a fire rating of at least H120. This will give a 2-hour protection period before having to evacuate or abandon the installation.

▷ Temporary Refuge (TR) buildings should be under positive atmospheric pressure using an airlock system.

▷ There should be more than one means of communicating to personnel specific instructions, such as what to do and where to go.

▷ The number of ways an alarm is conveyed to personnel should not be by siren alone i.e. a flashing beacon can be used for areas where a siren might not be heard.

▷ Multiple means of manually descending to sea level should be provided, e.g. knotted rope, sea ladder attached to the platform leg, scramble net, skyscape, etc.

▷ Appropriate personnel escape equipment should be available either in the accommodation area or on each escape route. This equipment should include Emergency Breathing Systems (EBSs), Emergency Life Support Apparatus (ELSA), smoke hoods, torches and flame retardant gloves.

When an incident occurs on an offshore installation and escape and evacuation is necessary, there are a number of means of leaving the installation safely. The primary method is by lifeboat. These can be launched by davit (a crane-like device which lowers the lifeboat to the sea) or by free-fall where the lifeboat is set at an angle on a launch ramp and allowed to fall into the sea when required.

Life rafts are another means of leaving the installation safely. However, they are not as efficient or as quick to escape in as lifeboats and should rank as a second choice means of escape.

Other means of descending to sea level in an emergency include by knotted rope, by sea ladder attached to the platform leg, by scramble net, or by 'skyscape' (a ladder-like escape device).

Escape can also be considered by helicopter. However, this is usually restricted to casualties who are not capable of leaving the installation unaided.

The platform's standby vessel will be available to respond to an escape and evacuation situation. As well as assisting with the evacuation from the platform, it can offer sea rescue with its Fast Rescue Craft (FRC) and receive launched survival craft (lifeboats and life rafts). If necessary it can also use its fire-fighting capabilities.

Roles and operation of fire teams onshore and offshore in upstream and downstream facilities

Onshore fire teams

When called upon to deal with an incident, the role of the on-site fire team is to control the situation as far as possible and make an evaluation to pass on to the emergency services when they arrive and take over command. Response teams are only expected to conduct search and rescue operations should personnel be unaccounted for, and are trained in the use of breathing apparatus.

Offshore fire teams

Offshore, the industry has moved to a position where on-board automatic, remote and fixed fire-fighting systems are expected to control, contain and bring to a satisfactory end almost any emergency. However, the platform's standby vessel may be able to assist with water cannon if this is deemed appropriate.

187

As with their onshore counterparts, offshore response teams now have the role of search and rescue for unaccounted-for personnel, and are trained in the use of breathing apparatus.

The offshore installation has a limited amount of external assistance it can call upon in an emergency, most of which has a significant time lag in how quickly it can respond. These services include:

- The platform Standby Vessel (SBV)
- Coastguard and search and rescue services (helicopter and lifeboat services)
- Commercial helicopters
- Supply vessels
- Shipping in the vicinity

Training and drills

Training

Training and drills, in relation to emergency response, are about ensuring everybody is in a state of preparedness and knows exactly what to do and what will be expected of them should an emergency arise.

- People in command should be competent. Competence can be defined as 'having sufficient training, knowledge and experience to undertake a task or duty safely and effectively'.
- There should be a sufficient number of competent people on the installation to undertake emergency duties and operate relevant equipment, including sufficient numbers of people to be in attendance at the helicopter landing area during helicopter movements.
- Everybody aboard an offshore installation should have undergone general training in emergencies, including training in personal survival, installation-specific induction training and training based on the emergency response plan.

Training specific to offshore installations

Everyone who is working offshore has to undergo general training in emergency situations. This is known officially as the Basic Offshore

Safety Induction and Emergency Training (BOSIET). This basic training covers:

▷ First aid
▷ Basic fire-fighting
▷ Self-rescue
▷ Helicopter safety and escape
▷ Sea survival

Beyond the basic course, which everybody has to undertake, more specific courses for designated personnel who will play a specific part in any emergency situation have been developed.

Drills

Drills, in relation to emergencies, are exercises which are undertaken to evaluate emergency plans and procedures under realistic conditions. They also serve the purpose of training personnel in the practical application of their roles and responsibilities. Consequently, they should be conducted as often as is deemed necessary in order to maintain competency levels.

External support agencies and resource liaison including municipal and offshore

Strategic command posts

When a major incident occurs, the emergency services will establish a set of strategic command posts to oversee and manage the situation. These are likely to be made up of:

▷ The main HQ commanders responsible for determining the best strategy for dealing with the incident.
▷ Strategic commanders are the next level down, possibly located at the rally point for the emergency services. These commanders are responsible for converting the strategy set out by the HQ based main commanders into a set of actions or tactics, to be implemented by the emergency teams at the scene.
▷ The final level of command is the supervisors in charge of the emergency workers at the scene. They will be in touch with

189

the strategic commanders to find out what strategy and/or tactics have been established as the best way to deal with the incident.

External support agencies

Other external assistance might include:

▶ Local councils
▶ The health and safety authority
▶ The agency responsible for the environment
▶ The water/drainage company
▶ The coastguard
▶ Air Sea Rescue
▶ Lifeboat rescue

Liaison with emergency services

In order to ensure any emergency is responded to effectively, it is essential that good channels of communication are established between the management of a facility and the emergency services. The groundwork for these channels will be set as part of the drills and exercises the parties jointly take part in, but over and above these activities, a constant liaison with the emergency services should be established and maintained.

When an emergency is responded to by the emergency services, there is a certain amount of information which will be required immediately. That information will include:

1 Contact point for the fire/police/ambulance liaison officer
2 Contact point for fire/police/ambulance incident commander
3 Rendezvous point for emergency services
4 Strategic response group member (planning co-ordinator or as otherwise delegated) with primary responsibility for managing the emergency at the site
5 Likely cause and effect of the emergency
6 Likely casualty status including potentials, how serious and their current location

190

7 Roll call results
8 Map of the site including floor plans, entry and exit points
9 Evacuation location
10 Outline of the local environment and surrounding risks (possibility
   of secondary incidents/contamination)
11 Utility shut-off points
12 Availability of CCTV
13 Press liaison details
14 Welfare arrangements
15 Traffic control points and likely impacts on the surrounding area

Sub-element 4.2: Emergency response

Learning outcome

Outline the principles, procedures and resources for effective
emergency response.

✎ Revision exercise

Write your answers to the questions below on a separate sheet
of paper without referring to the information in this book in the
first instance. Once you have answered all of the questions, you
can refer back to the revision guide to compare your answers and
this will give you an indication of how much knowledge you have
been able to absorb or whether you need to revise this section
further.

Q1 **Outline** what the general contents of an emergency plan
    include.
Q2 **Explain** what a 'fire and explosion strategy' is.
Q3 **Outline SIX** measures which will be considered within a fire
    and explosion strategy.
Q4 **Explain** the difference between an alarm used on onshore
    facilities and an alarm used on offshore facilities.
Q5 **Explain** what is meant by the phrase 'medical triage'.

Q6   **Outline SIX** factors which will enhance an escape and evacuation route from an offshore facility.

Q7   **Give THREE** examples of emergency evacuation from an offshore facility.

Q8   **Describe** what the underlying principle is with regard to training and drills for emergency response.

Q9   **Explain** the purpose of undertaking drills in relation to emergencies.

Q10   **Explain** why it is important that the management of facilities liaise with emergency services.

Q11   **Outline SIX** pieces of information which will be required by the emergency services in the event of an emergency.

Element 5

CHAPTERS 19–20

Sub-element 5.1:
Marine transport

Learning outcome

Identify the main hazards of, and suitable controls for, marine transport in the oil and gas industries.

Key revision points

Hazards of vessels and working over water ☐

Loading and unloading of vessels at marine terminals ☐

Control of marine operations, certification of vessels, inspection and approvals ☐

Roles and responsibilities of marine co-ordinators, masters and crew ☐

Personnel transfers and boarding arrangements ☐

Personal protective equipment suitability ☐

Diver operations ☐

Hazards of vessels and working over water

Introduction

Shipping plays an essential part in providing necessary support and services to offshore installations. Consequently, marine activities present many unique risks and hazards which require special consideration in order to control them.

As dangerous as passing vessels are, the majority of collisions with offshore installations involve attendant vessels. Attendant vessels cause around ten times more severe damage collisions than passing vessels and can result in catastrophic losses.

In order to minimize the risks associated with attendant vessels colliding with installations, support vessels should generally work on the lee side (downwind) of the installation. If support vessels have to work with the rig on a side other than the lee side, a full risk assessment should always be conducted prior to the work commencing.

Exclusion zone

The immediate area around an installation is regarded as a major risk area with potentially severe consequences for any incident that happens within it. Consequently, it is designated as an exclusion zone with restrictions on which vessels are allowed to enter the zone.

The zone extends for 500 metres around the installation and is constantly monitored by radar, as well as being patrolled by the platform's Standby Vessel (SBV) or its emergency response and rescue vessel, either of which is in close communication with the platform's Central Control Room (CCR). Any vessel wishing to enter the exclusion zone must seek permission from the central control room and the patrol vessel before doing so. If a vessel tries to enter the zone without permission, it will be warned off by the patrol vessel.

Vessel hazards

Other hazards associated with vessels include:

▶ Breakdown, loss of power or loss of steering. This can lead to drifting, collision, running aground, etc.
▶ Anchoring over pipelines, wells and submerged cables. This can lead to damage or rupture of pipelines, wells or cables.
▶ Explosion during loading/unloading operations.
▶ Pollution – spillage, leakage, etc.
▶ Striking the installation (e.g. by the platform supply vessel in adverse weather).
▶ Man Overboard (MOB). The personal hazards associated with someone who falls into the water include:
  ▷ drowning
  ▷ hypothermia
  ▷ being struck by debris or vessel
  ▷ becoming entrapped by debris.

Types of vessels and activities associated with the oil and gas industry

Types of vessels – platform support vessel

The Platform Support Vessel (PSV) acts as a shuttle between the offshore platform and the mainland, bringing all the goods, materials and spare parts required to keep the platform operational. It also takes back to the mainland any waste material and equipment in need of repair.

The PSV has to work in close proximity to the platform in order to unload and load its cargo.

Hazards include:

▶ Collision between the PSV and the platform.
▶ Lifting hazards from crane operations. This includes shock loading, which is when craning operations are disrupted by the swell of the sea lifting the PSV.
▶ Dropped objects from the platform to the PSV.

197

Types of vessels – floating assets

A Floating Production, Storage and Offloading (FPSO) unit is a vessel used for processing and storing oil and gas extracted from the well head. This is then offloaded at regular intervals to tankers for transportation to onshore terminals/refineries.

Offloading operations

The transfer of hydrocarbon product from the FPSO unit to the tanker is a hazardous operation, with the main hazard being that the two vessels may come together as the tanker manoeuvres into position. In order to reduce this risk of collision, Yokohama fenders are placed between the two vessels so they do not make contact with each other.

Single buoy mooring

There are many remotely situated sea-based well heads throughout the world which only have a buoy connected to them. These are known as Single Buoy Moorings (SBMs) or Single Point Moorings (SPMs). These buoys act as a mooring point for tankers and have within them a product transfer system which facilitates the transfer of hydrocarbon product to the tanker.

The main hazard associated with this operation is the potential for the tanker and buoy to come together. The main control for this hazard is to have a support vessel secured to the stern of the tanker to hold the tanker off the buoy.

Loading and unloading of vessels at marine terminals

Transfer of material between marine vessels and tanks

Before loading operations commence, a loading plan is formulated. The filling of tanks must be monitored and procedures must be followed which minimize the risks associated with these operations. These include:

▷ The vessel must be securely moored with sufficient mooring scope to ensure it does not range along or away from the berth.

▷ The mooring scope must also take into account tidal rise and fall, river currents and the possible effects of passing ships.

▷ Ensuring hoses are suitable for the product being discharged and the operating pressures they will be subjected to.

▷ Ensuring the connections of pipes and hoses to be used in the transfer operation are secure.

▷ Positioning drip trays beneath all connections and ensuring there is close monitoring of connections during transfer operations.

▷ Deploying fire wires on the vessel to give tugs a means of moving the vessel away from its berth quickly if an emergency arises.

▷ Reducing the risk of static electrical charges occurring on board ship by ensuring all metal objects are bonded to the ship.

▷ Due to the possible differences in electrical potential between the ship and the berth, there is a risk of electrical arcing at the manifold during connection and disconnection of the shore hose or loading arm. To protect against this risk, there should be a means of electrical isolation at the ship/shore interface.

▷ Fire control measures, such as fire-fighting equipment, should be made ready before transfer commences.

▷ Agreement should be reached between ship and shore on a discharge plan which ensures the vessel is not subject to undue internal stresses as the cargo is discharged. See the Widdy Island disaster case study.

▷ The control room should monitor flow rates and quantities. This includes alarm systems to indicate when tanks are nearing their filling point. There are also sensors indicating the trim of the marine vessel so that adjustments can be made to the ballast of the vessel as required.

▷ All doors and windows aboard the vessel and in buildings at the terminal to be closed. This is to ensure there is no ingress of flammable vapour which might build up with the potential of causing an explosion.

▷ Adequate venting arrangements should be in place to ensure vapour is dispersed properly and safely. These will include

199

monitoring wind direction and strength. Low wind speed can be an added hazard, as the dispersion of vapour in these conditions is minimal and it can build up in dangerous quantities without being apparent.

▷ Venting arrangements should be made for both the recipient tank and the donor tank. The donor tank will require a volume of air, or more likely inert gas, to replace the volume of product transferred.

▷ Once discharge commences, the vessel must be kept within the operating envelope (limits) of the oil loading arms.

Control of marine operations, certification of vessels, inspection and approvals

Certification of vessels

The role of the International Maritime Organization (IMO) is to develop international conventions that set out the minimum acceptable standards for the maritime industry. These standards include such things as the construction of the vessel, the safety equipment it must carry, and the training and certification of crew.

All vessels are required to be registered under a 'flag state' when they are first built, which means the owners of the vessel agree to comply with the maritime regulations of the flag state.

During the initial registration process, the vessel is inspected by the flag state inspection team and, provided it meets the required standards, it will be issued with a number of certificates, each relating to various aspects of the vessel. These certificates are renewed at various intervals on satisfactorily passing a re-inspection.

Roles and responsibilities of marine co-ordinators, masters and crew

Roles and responsibilities of masters and crew

On any seagoing vessel, the person responsible for the safety of the vessel and all those aboard is the master. The master has the absolute right – and duty – to make the final decision on matters affecting the vessel and those on board.

The size and structure of the crew will depend on a number of factors, including the size of the vessel, the type of cargo, the requirements of the owner, etc. A typical crew list for a seagoing tanker is set out below.

Table 19.1 – Roles, responsibilities and typical numbers of ship's crew

Rank	Responsibility	Reports to	Numbers on board
Master	Overall command	Head office	1
Chief officer/ mate	Maintenance of deck. Cargo ops. Bridge watch at sea.	Master	1
Second mate	Navigator. Deck watch in port. Bridge watch at sea.	Master. Cooperates with mate.	1
Third mate	Maintains life-saving equipment and fire-fighting equipment. Deck watch in port. Bridge watch at sea.	Master. Cooperates with mate.	1
Bosun/CPO	Deck crew supervisor	Mate	1
Able bodied seaman/SG1	Skilled deck worker. Rigger. Helmsman.	Bosun	4
Deckhand	Deck worker in training	Bosun	2
Chief engineer	Overall technical maintenance of vessel	Master	1
Second engineer	Engine room maintenance	Chief engineer	1
Third engineer	Power generators	Chief engineer. Cooperates with second engineer.	1
Fourth engineer	Anything to do with fuel	Chief engineer. Cooperates with second engineer.	1
Fitter	Technical maintenance supervisor	Second engineer	1

Rank	Responsibility	Reports to	Numbers on board
Pumpman	Technical deck maintenance	Mate and chief engineer	1
Motorman	General engine room maintenance	Fitter	3
Chief steward	In charge of catering and housekeeping	Master	1
Cook		Chief steward	2
Stewards	Housekeeping and waiting on tables	Chief steward	2
Cargo engineer	On Liquefied Natural Gas (LNG) carriers, refrigeration, etc. for cargo	Chief engineer	1

Source: Adapted from www.ilo.org/dyn/normlex/en/f?p=1000:53:0::NO:53:P53_
FILE_ID:3130434; and www.maritime-transport.net/mtso/downloads/Public_
Information/MTCP_report_safe_manning_level_study.pdf

Roles and responsibilities of marine co-ordinator

A marine co-ordinator can be based on board a Floating Production, Storage and Offloading (FPSO) unit or at a terminal. The role involves being responsible for the co-ordination, testing and maintenance of all marine systems and equipment. He/she is also responsible for ensuring all marine activities, procedures and guidance are in compliance with current legislation, codes and standards.

The marine co-ordinator will also lead on cargo transfer operations.

Personnel transfers and boarding arrangements

Gangway boarding

This is the main and safest method of boarding a vessel. The gangway should have basic safety features such as handrails, non-slip treads and a suspended cargo net beneath the gangway. If the vessel is a tanker, the Emergency Shutdown (ESD) facility is also positioned adjacent to the gangway.

The gangway will always be manned by a watchman who will be in radio contact with the officer of the watch. The watchman's duties include implementing security procedures as well as keeping and up-dating a Personnel On Board (POB) list. This is so that if an emergency arises, there is accurate knowledge of who needs to be accounted for.

At the shore side of the gangway there will also be a fire plan so that, in case of emergency, shore responders can access a plan of the ship.

Accommodation ladder boarding

An accommodation ladder is an access ladder which is a permanent feature of the ship and is connected to a ship's side. Its elevation can be adjusted according to the requirements of those wishing to board or leave the vessel.

Pilot ladder boarding

A pilot ladder is for use by pilots to board and leave a ship. It should not be used by any other personnel for transferring between vessels as it requires training and experience to undertake the transfer safely. If personnel need to board or leave a vessel from an attendant vessel, this should be done via the accommodation ladder.

The main hazard with pilot ladder transfer is judging the swell of the sea whilst stepping from the ladder onto the other vessel. It also requires an experienced coxswain to ensure the pilot boat does not foul the ladder and cause undue strain, which might lead to the pilot ladder being torn from its deck fixing points.

Other types of personnel transfer

There are two other types of personnel transfer which are occasionally used.

The first is the basket transfer, sometimes known as the Billy Pugh. The basket is hoisted and lowered by a crane. The personnel being

203

hoisted stand on the basket's perimeter, lean in towards the net and hold tight before being lifted.

The second type of basket transfer is known as the 'Frog'. Unlike the Billy Pugh, the people being transferred climb inside the Frog and get strapped into it.

Personal protective equipment suitability

Personal Protective Equipment (PPE) – specific requirements for marine use

- Lifejackets should be used when working over the side or on deck in heavy weather.
- Wet weather and/or cold weather clothing should be capable of providing adequate protection. Temperatures as low as −25°C can be experienced and, when coupled with a wind chill factor, can present extreme conditions.
- Orinasal masks fitted with suitable filter cartridges should be used on liquefied petroleum gas and liquefied natural gas vessels.
- Emergency Life Support Apparatus (ELSA) should be installed at strategic places to aid escape from leaking gas on LPG and LNG carriers.

Diver operations

Diving project plan

Once the extent of the work that requires diver operations to complete has been established and a risk assessment has been carried out covering the work, a diving project plan can be formulated.

The project plan will show how and to what extent the work will be split up into separate dive operations; it will show how the hazards identified in the risk assessment will be controlled; and it will include emergency and contingency plans.

Diving contractor's responsibilities

The diving contractor's general responsibilities are to ensure that:

- The diving project is properly and safely managed.
- Risk assessments have been carried out.
- The place from which the diving is to be carried out is suitable and safe.
- A suitable diving project plan is prepared which includes emergency and contingency plans.
- The supervisor and dive team are fully briefed on the project and are aware of the contents of the diving project plan.
- There are sufficient personnel in the dive team to enable the diving project to be carried out safely.
- The personnel are qualified and competent.
- Supervisors are appointed in writing and the extent of their control fully documented.
- A suitable mobilization and familiarization programme is completed by all the members of the dive team. Other personnel involved in the diving project, for example ship's crew, may also need to complete the programme.
- Adequate arrangements exist for first-aid and medical treatment.
- Suitable and sufficient plant is provided and that it is correctly certified and maintained.
- The divers are medically fit to dive.
- Diving project records are kept containing the required details of the diving project.
- There is a clear reporting and responsibility structure laid down in writing.

Diving project – further control measures

Over and above the responsibilities of the diving contractor, there are other basic control measures which should be applied to all diving operations. These include:

- Any dive vessel must remain on station throughout the dive operation.

205

▶ Other subsea work may need to be suspended whilst divers are at work.

▶ All work being conducted above the divers (e.g. construction) should be suspended during the dive operation.

Sub-element 5.1: Marine transport

Learning outcome

Identify the main hazards of, and suitable controls for, marine transport in the oil and gas industries.

Revision exercise

Write your answers to the questions below on a separate sheet of paper without referring to the information in this book in the first instance. Once you have answered all of the questions, you can refer back to the revision guide to compare your answers and this will give you an indication of how much knowledge you have been able to absorb or whether you need to revise this section further.

Q1 **Describe** what an 'exclusion zone' around an offshore facility is.

Q2 **Outline THREE** hazards associated with vessels in relation to offshore facilities.

Q3 **Explain** what 'single buoy mooring' is and its purpose.

Q4 The loading and unloading of vessels at marine terminals can be a hazardous event. **Outline SIX** procedures or controls which will minimize the risks associated with these operations.

Q5 **Explain** what the process of 'vessel certification' involves.

Q6 **Outline** the responsibilities of the master of a seagoing vessel.

Q7 **Explain** the difference between gangway boarding of a vessel and accommodation ladder boarding.

Q8 **Explain** what a 'diving project plan' is.

CHAPTER 20

Element 5

Sub-element 5.2:
Land transport

Learning outcome

Identify the main hazards of, and suitable controls for, land transport in the oil and gas industries.

🔑 **Key revision points**

Road tankers	☐
Traffic management	☐
Rail	☐

Road tankers

Road transportation of hazardous materials

The first step in the safe carriage of hazardous material is to be able to:

▷ Identify exactly what it is that is being carried
▷ What the precise hazards are that this product presents
▷ The best means of controlling those hazards

The UN classification and labelling system for the transportation of dangerous goods, which is now universally accepted, is the means by which this information is made available on all vehicles carrying dangerous goods.

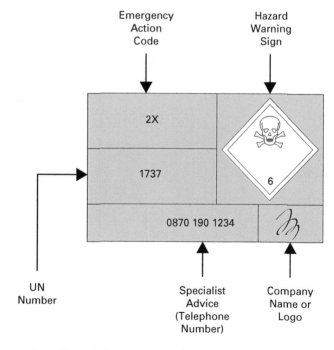

Figure 20.1   Hazard signboard panel

Source: Wise Global Training.

The system uses a standard signboard fixed to the vehicle in a designated position or positions. On this signboard are a number of internationally accepted series of codes and symbols to show what is being transported, the hazards it represents and the correct preventative actions to be taken when required.

General principles for avoiding plant being struck by vehicles

▷ Traffic routes should be wide enough for the safe movement of the largest vehicle permitted to use them (including visiting vehicles).
▷ Traffic routes should have enough height clearance for the tallest vehicle permitted to use them (including visiting vehicles).
▷ Potentially dangerous obstructions, such as overhead electric cables, or pipes containing hazardous chemicals, need to be protected using goal posts, height gauge posts or barriers.
▷ Traffic routes should be planned to give the safest routes between calling places.
▷ Routes should avoid passing close to such things as unprotected fuel or chemical tanks or pipelines.
▷ All potentially vulnerable plant should be protected from errant vehicles by collision barriers.
▷ All routes should be well lit.
▷ Any hazardous sections of the route, such as sharp bends or adverse cambers, should be clearly signed.
▷ A person should be appointed to be responsible for, and oversee, site traffic movements on site.
▷ Drivers should be trained and authorized to drive vehicles on site.
▷ Visiting drivers should be briefed on site traffic movement rules.
▷ Reversing of vehicles should be avoided or controlled.

Driver training for transportation of dangerous goods

The main objectives of driver training, in relation to carrying dangerous goods, are to ensure drivers:

209

▷ Are aware of the hazards arising when they are driving a vehicle which is carrying dangerous goods

▷ Know what steps to take in order to reduce the likelihood of an incident taking place

▷ Know what necessary measures they need to take when driving a vehicle which is carrying dangerous goods with regard to ensuring their own safety, that of the public and the environment. Also, if an incident does occur, drivers should know how to limit the effects of that incident

▷ Have practical experience of what actions they will need to take in the case of an incident occurring

Loading and discharging arrangements

The following control measures should be implemented when the loading and unloading of hydrocarbon materials takes place.

▷ The area designated for loading and unloading should be situated away from general traffic routes. It should also be situated on level ground.

▷ There should be sufficient space to allow the largest planned-for vehicles to easily manoeuvre into and out of the loading/unloading area.

▷ Loading/unloading areas should be adequately lit when in use.

▷ A system should be implemented that ensures a vehicle cannot be driven away from the loading/unloading point before being authorized to do so.

▷ When the loading/unloading operation has been completed, drivers must ensure all tank openings, including valves and caps, are closed before starting their journey.

▷ No tank should be overfilled as most tanks require room for expansion of the liquid.

▷ When filling tanks, the pressure in the tank must be monitored to ensure it does not exceed its maximum working pressure. Tanks should be fitted with pressure relief valves.

▷ When discharging tanks, the pressure in the tank must be monitored to ensure a vacuum is not created. Tanks should be fitted with vacuum breaker valves.

210

▶ All external vents should be fitted with flame arrestors.

▶ The rate of filling or discharge must be limited. This is to reduce the risk of static electricity build-up, which can be caused by splashing and liquid being circulated unduly.

▶ Suitable drip trays should be placed beneath hose connection points when loading/unloading operations are being set up or are being ended.

▶ Where product with a flash point of 60°C or less is being loaded/unloaded, a bonding wire should be connected between the vehicle and an earthing point before loading/unloading commences.

▶ A no smoking policy should be established and maintained on site.

▶ There should be two opposing emergency exits from the loading/unloading area.

▶ Exits should be clearly marked and open outwards.

▶ Vapours which are displaced during the transfer operation should be returned to the donor tank via a vapour-tight connection line.

▶ The vapour return line should have a different connection fitting compared with the product transfer hose. This is to ensure there can be no misconnection.

▶ The vapour return line should be connected before the product transfer hose is connected.

▶ There should be a device on the vehicle which locks the brakes in the 'on' position when the vapour recovery line is connected.

▶ A competent person should be given the responsibility to monitor all the hose connections during loading/unloading operations.

▶ Any uncontrolled release of vapour should be recorded in the vehicle log book and reported to the authorities.

▶ There should be a pre-formulated spillage plan ready to deal with any spillages and a spillage kit at the ready. This kit should include bunding.

Traffic management

Vehicle movements on site

Vehicle movements can be managed on site by implementing the following controls:

▶ A person should be appointed to be responsible for, and oversee site traffic movements on site.
▶ All routes should be well lit.
▶ Any hazardous sections of the route, such as sharp bends or adverse cambers, should be clearly signed.
▶ Drivers should be trained and authorized to drive vehicles on site.
▶ Visiting drivers should be briefed on site traffic movement rules.
▶ Reversing of vehicles should be avoided or controlled.
▶ A contingency plan and the resources to implement it should be in place in case of adverse weather conditions such as snow and ice.
▶ Periodic surveys and safety tours should be conducted to ensure traffic rules are being complied with.

Vehicle movements off site

Prior to leaving the site, check that:

▶ All hoses are secure.
▶ There are no leaks from the hoses.
▶ Blanking caps are fitted.
▶ The load is not leaking.
▶ The load is not overheating.
▶ For liquefied petroleum gas tanks, the pressure is within prescribed limits.
▶ The brakes and tyres on the vehicle are in working order.
▶ All documents are in order and available.
▶ The markings are correct and in place.

When planning the route, employ the following controls:

▶ Avoid built up areas.

▶ Avoid roads with low bridges.
▶ Obey any restrictions relating to the transport of dangerous goods.

During the journey check that:

▶ The load is still secure.
▶ There are no leaks.
▶ There is no overheating on the vehicle that could lead to fire, i.e. tyres, bearings, brakes, etc.
▶ Any control temperature is not exceeded.
▶ Markings are still in place, clean and visible.

Driving techniques which:

▶ Maintain concentration.
▶ Try to anticipate situations wherever possible.
▶ Plan ahead.
▶ Drive defensively.
▶ Keep to the speed limit and drive at a safe speed especially at roundabouts, motorway slip roads and site entrances.
▶ Always allow a margin of safety.

Rail

The international carriage of dangerous goods by rail within Europe is governed by Annex I of the 'Convention Concerning International Carriage by Rail'. This is known by the letters RID. However, the principles of these regulations form a sound basis for the transport of hydrocarbon products by rail anywhere in the world.

In order to transport hydrocarbon products by rail, a number of duties must be complied with. These include:

▶ Any dangerous goods being transported by rail must be clearly marked and labelled with their name, description and UN number, just as with road transport.
▶ Only carriers with the appropriate resources and experience should be engaged to carry hydrocarbon products.

▷ All temporary storage of rail traffic carrying hydrocarbon products should be secure.

▷ Rail staff engaged in the transportation of hydrocarbon products should undergo awareness training programmes.

▷ There should be an emergency plan in place where hydrocarbon products are involved.

Rail loading/unloading operations present additional hazards that are unique to rail. Precautions against these are:

▷ The movement of trains through entrances and exits of rail sidings should be supervised by a competent person.

▷ Sidings should be isolated from any main line.

▷ During loading/unloading operations, warning signs should be displayed on the train if open to access. These will include:

  ▷ red flag during the day
  ▷ red light at night
  ▷ a warning sign that the rail cars are connected.

▷ Loading/unloading operations must be monitored throughout the operation.

▷ Rail cars which have been disconnected from the locomotive must be prevented from moving.

▷ There should be a 15-metre exclusion zone around any loading/unloading point. This exclusion zone will prohibit any potential source of ignition.

▷ Where tools have to be used, these should be non-sparking tools.

▷ Prior to loading/unloading operations commencing, a system of vapour control should be established.

▷ The closure of all foot valves, lids and the removal of hoses, etc. must be overseen by a competent person.

▷ If rail tankers are fitted with product heaters, these must *not* be used if flammable vapours are present during loading and discharge operations.

▷ When work involves working on top of rail cars, working-at-height regulations should be adhered to.

▷ Weather conditions (the potential for lightning) should be taken into consideration prior to the commencement of loading/unloading.

Sub-element 5.2: Land transport

Learning outcome

Identify the main hazards of, and suitable controls for, land transport in the oil and gas industries.

✎ Revision exercise

Write your answers to the questions below on a separate sheet of paper without referring to the information in this book in the first instance. Once you have answered all of the questions, you can refer back to the revision guide to compare your answers and this will give you an indication of how much knowledge you have been able to absorb or whether you need to revise this section further.

Q1 **Explain** what purpose the UN classification and labelling system for the transportation of dangerous goods serves.

Q2 **Outline FIVE** measures which can be applied to avoid plant being struck by vehicles.

Q3 **Explain** the purpose of driver training for the transportation of dangerous goods.

Q4 **Outline SIX** control measures which can be implemented when loading and unloading of hydrocarbon materials takes place.

Q5 **Give FOUR** control measures which can be implemented to manage vehicle movements safely on site.

Q6 **Give FIVE** examples of checks which should be carried out prior to a vehicle carrying hydrocarbon material leaving site.

Q7 **Outline FOUR** control measures which are specifically applicable to the loading/unloading of hydrocarbon materials from rail vehicles.

Appendix

Answers to Revision Questions

Revision questions and suggested outline answers for Sub-element 1.1

Q1 Outline FIVE reasons for investigating accidents and incidents. Include in your answer at least one legal reason and two financial reasons.

Q1   Answer – five of the following

Legal reasons for investigating accidents and incidents include:

▶ To demonstrate that the company is meeting its legal requirements.

▶ Employers should be able to make available information in case those involved in the accident decide to take legal action.

▶ A company can demonstrate to the courts their commitment and positive attitude to health and safety.

Financial reasons for investigating accidents and incidents include:

▶ Information forthcoming from an accident investigation provided to an insurance company may well assist in the event of a claim.

▶ The outcome of an investigation could prevent a recurrence.

▶ The company can avoid business losses by preventing further similar incidents.

▶ Other costs saved might include the cost of legal action which may be taken against the company; increased insurance premiums; loss of business due to a bad reputation resulting in lost orders.

Q2 **Identify the kind of persons you would expect to be involved in a team set up to investigate an accident/ incident.**

Q2   Answer

▶ Operations team leader
▶ In the case of an offshore installation, a field or platform safety officer
▶ In the case of an onshore installation, a senior onshore manager
▶ Safety representatives
▶ Area authorities (the person responsible for the area where the incident took place)
▶ Specialist inspectors
▶ If it's a drilling rig, a tool pusher

Q3 **Identify the FOUR stages in the process of investigating accidents and incidents.**

Q3   Answer

▶ **Step one** – gathering the information
▶ **Step two** – analysing the information
▶ **Step three** – identifying the required risk control measures
▶ **Step four** – formulation of the action plan and its implementation

Q4 **Identify SIX observational skills or techniques used in accident/incident investigations.**

Q4   Answer – six of the following

▶ Knowledge of the workplace and procedures
▶ Keeping a systematic record of observations
▶ Take time to observe the whole scene
▶ Being alert to possible changes to the accident scene by those who may have a motive to correct unsafe practices
▶ Looking Above, Below, Behind, Inside – ABBI
▶ Being inquisitive and questioning employees to determine risks
▶ Using all senses including smell, sight, touch and hearing
▶ Having an open mind and looking for solutions

▷ Identifying, recording and feeding back good performance as well as bad

▷ Using an interviewing style which does not reflect a blame culture

▷ Asking questions in a way which does not make the interviewee feel intimidated or uncomfortable

▷ Conducting the interview in non-intimidating surroundings

▷ Encouraging witnesses to speak openly in their own words without using technical jargon

▷ Promoting a positive attitude to finding the reasons for the incident rather than apportioning blame

▷ Interviewing witnesses separately and in private to prevent them from influencing each other's accounts

▷ Providing a summary of what the witness said in order that they can ensure that everything has been understood correctly and that the interviewer has not misinterpreted the account

Q5 Give SIX sources of information relating to accident/ incident investigations.

Q5   Answer – six of the following

▷ Victim statements

▷ Witness statements

▷ Plans and diagrams

▷ CCTV coverage

▷ Process drawings, sketches, measurements, photographs

▷ Check sheets, permits-to-work records, method statements

▷ Details of the environmental conditions at the time

▷ Written instructions, procedures and risk assessments which should have been in operation and followed

▷ Previous accident records

▷ Information from health and safety meetings

▷ Technical information/guidance/toolbox talk sheets

▷ Manufacturers' instructions

▷ Risk assessments

▷ Training records

▷ Logs

▷ Instrument readouts and records

▷ Opinions, experiences, observations

Q6   In relation to causes of accidents, explain what is meant by:

(a) Immediate causes

(b) Underlying causes

(c) Root causes

▶ Immediate causes are generally unsafe acts and/or conditions

▶ Underlying causes are generally procedural failures

▶ Root causes are generally management system failures

Q7   Give the hierarchy of risk control.

1   Eliminate the risk

2   Replace the risk for something safer

3   Apply engineering controls

4   Apply administrative controls

5   Use Personal Protective Equipment (PPE)

Q8   Lessons learnt from major incidents can be disseminated both locally and more widely. Outline who might benefit from lessons learnt:

(a) Locally

(b) More widely

Learning lessons locally involves

▶ The management of the organization

▶ The regulatory bodies

▶ The operators of the systems and procedures involved in the incident

▶ The incident investigators generally

Learning lessons more widely involves

▶ The lessons being disseminated widely throughout other organizations and the industry in general

Revision questions and suggested outline answers for Sub-element 1.2

Q1 Outline the features of the following:

(a) Flash point

(b) Vapour density

(c) Vapour pressure

Q1 Answer

The flash point of a volatile liquid is the lowest temperature at which it can vaporize to form an ignitable mixture when mixed with air.

Vapour density is the measurement of how dense a vapour is in comparison with air.

Vapour pressure is the process of evaporation of the molecules on the surface of a liquid.

When the energy within these molecules is sufficient for those molecules to escape, they do so in the form of a vapour.

Q2 Explain the difference between 'flammable', 'highly flammable' and 'extremely flammable'.

Q2 Answer

'Flammable' describes a product which is easily ignitable and capable of burning rapidly. In the UK a flammable liquid is defined as a liquid that has a flash point of between 21°C and 55°C.

'Highly flammable' describes a product which has a flash point below 21°C but which is not defined as extremely flammable.

'Extremely flammable' describes a product which has a flash point lower than 0°C and a boiling point of 35°C or lower.

Q3 Explain what the 'lower flammable limit' and 'upper flammable limit' of a product is.

Q3 Answer

The lower flammable limit is the lowest concentration of a gas or vapour in air which is capable of being ignited. The upper flammable limit is the highest concentration of a gas or vapour in air which is capable of being ignited.

Q4   Toxicity can be 'acute' or chronic'. Explain the difference between the two terms.

Q4   Answer

'Acute toxicity' is a term which describes the effect a substance has on a person after either a single exposure or from several exposures within a short space of time (e.g. 24 hours or less).

'Chronic toxicity' is a term which describes the effects a substance has on a person after many exposures over a longer period of time (e.g. months or years).

Q5   Some substances are described as 'skin irritants'. Explain what a 'skin irritant' is.

Q5   Answer

A skin irritant is a chemical which is not corrosive but which causes a reversible inflammatory effect on living tissue by chemical action at the site of contact.

Q6   Explain what is meant when a substance is said to have 'carcinogenic properties'.

Q6   Answer

A carcinogen is a substance that can cause, or aggravate, cancer.

Q7   Outline hazards associated with:

(a) hydrogen gas
(b) hydrogen sulphide
(c) methane
(d) Liquefied Petroleum Gas (LPG)
(e) Liquefied Natural Gas (LNG)
(f) nitrogen
(g) oxygen
(h) micro-biocides
(i) refrigerants
(j) steam
(k) mercaptans

Q7   Answer

Hydrogen is a highly flammable gas when it is mixed with air. It also burns with an invisible flame.

Hydrogen sulphide is a toxic, corrosive and flammable gas; it is heavier than air, tends to drift in low lying areas such as pits, cellars and drains, and is difficult to disperse.

Methane is a flammable gas and can be explosive in certain concentrations. It can also cause asphyxiation if the concentration is high enough.

LPG can cause a massive vapour cloud from a relatively small amount of liquid when that liquid is released into the air, with the potential to cause a Boiling Liquid Expanding Vapour Explosion (BLEVE). It can also cause asphyxiation, cold burns to the skin on contact, brittle fracture to carbon steel on contact and environmental damage.

LNG can cause asphyxiation, cold burns to the skin on contact and brittle fracture to carbon steel on contact.

Nitrogen has the potential to cause asphyxiation when it is used in confined spaces to displace oxygen.

Oxygen has the potential to cause asphyxiation when it displaces carbon dioxide and with it the stimulation to breathe. It allows a fire to burn and it can cause rusting.

Micro-biocides are classed as irritants to skin and eyes on contact as well as being toxic if ingested.

Refrigerants can cause injury from components or material ejected by the high pressure escape, frostbite, asphyxiation, possible explosion and possible secondary toxic gases if refrigerant gases burn. As refrigerant gases are heavier than air they will tend to drift in low lying areas such as pits, cellars and drains, and are difficult to disperse.

Steam has the potential to cause thermal shock and can cause burns to skin. High pressure steam can cause failure of components.

Some mercaptans are harmful. For example methyl mercaptan is a respiratory irritant and chronic exposure may cause lung damage. It is also a skin and eye irritant and it can depress the central nervous system. It also has a flashpoint of −18°C.

Q8   Outline hazards and control measures associated with:
(a) drilling muds
(b) low specific activity sludges

Q8   Answer

When muds come to the surface there may be natural gases or other flammable materials which have combined with the mud during the drilling operation. Consequently, there is risk of fire or explosion should these gases find a source of ignition.

Control measures for drilling muds include applying safe working procedures, installing monitoring sensors, using equipment and wiring which has been certified as explosion-proof.

Hazards from the sludges include the inhalation and ingestion of radionuclides, especially dust and fumes, skin irritant, direct radiation potential causing carcinogenic problems, and environmental pollution.

Control measures for sludges include the provision of ventilation equipment to control dusts and fumes, the use of wet methods of working and good housekeeping to reduce the amount of dust in the atmosphere, having equipment in place to collect sludge instead of using manual means, diluting sludge with water, the use of permit-to-work systems, the provision of training and awareness programmes, the provision of a health surveillance programme to monitor the health of employees, and the use of respiratory protective equipment specifically to protect against exposure to airborne radioactivity.

Revision questions and suggested outline answers for Sub-element 1.3

Q1   Give the five steps in a risk assessment process.

Q1   Answer

1   Identify the hazards
2   Decide who might be harmed and how
3   Evaluate the risks and decide on precautions
4   Record the findings and implement them
5   Review the assessment on a regular basis and update if necessary

Q2 Give the hierarchy of risk control measures.

Q2 Answer
1. Elimination
2. Substitution
3. Engineering controls
4. Administrative controls
5. Personal protective equipment

Q3 Explain what the difference is between a 'qualitative risk assessment', a 'semi-quantitative' risk assessment and a 'quantitative' risk assessment.

Q3 Answer

Qualitative risk assessment is based on the conclusions reached by the assessor using his/her expert knowledge and experience to judge whether current risk control measures are effective and adequate.

Semi-quantitative risk assessment involves applying a numerical value to the degree of severity and the likelihood of a particular event if it occurred.

Quantitative risk assessment involves using special quantitative tools and techniques in order to identify hazards and to give an estimate of the severity of the consequences and the likelihood of the hazards being realized.

Q4 Explain what a 'risk rating matrix' is.

Q4 Answer

A risk rating matrix is a table which uses a numerical value for the risk of an event happening and a numerical value of the consequences. When multiplied together, the resulting value gives an evaluation of the incident.

Q5 Explain what the purpose of a Hazard Identification Study (HAZID) is.

Q5 Answer

A HAZID is a tool for identifying hazards. It is normally a qualitative risk assessment and is judgement based. It is usually undertaken by a team of people who are selected for their particular knowledge, experience or expertise. The team compiles a list of hazards in order

225

to conduct a qualitative evaluation of how significant the hazards are and how to reduce the risks associated with them.

Q6 Explain what the purpose of a Hazard and Operability Study (HAZOP) is.

Q6 Answer

HAZOP is a tool which is used to systematically examine every part of a process or operation in order to find out how deviations from the normally intended operation of a process can happen and whether further control measures are required to prevent the hazards from happening.

Q7 Identify the kind of persons you would expect to be part of a HAZOP team.

Q7 Answer

- Chairperson
- Design engineer
- Process engineer
- Electrical engineer
- Instrument engineer
- Operations engineer

Q8 Explain what a Failure Modes and Effects and Critical Analysis (FMECA) study is.

Q8 Answer

FMECA is a method of systematically identifying the failure modes of an electrical or mechanical system. One or two people examine each component of the system in turn and evaluate the effects and the degree of importance if that component should fail.

Q9 Explain what is meant by 'As Low As Reasonably Practicable' (ALARP).

Q9 Answer

ALARP means that employers should adopt appropriate safety measures unless the cost (in terms of money, time or trouble) is grossly disproportionate to the risk reduction.

Q10 The management of major incident risks should take a hierarchical approach. Give that hierarchical approach.

Q10 Answer

1 Elimination and minimization of hazards
2 Prevention by reducing the likelihood of a major incident
3 Detection via the warning and alarm systems
4 Control by limiting the scale, intensity and/or the duration of an incident
5 Mitigation of consequences

Q11 Outline the main principles for achieving an inherent safe design.

Q11 Answer

▶ Minimizing the amount of hazardous material present at any one time
▶ Replacing hazardous materials with less hazardous materials
▶ Moderating the effect a material or process might have
▶ Designing out problems rather than adding features to deal with problems
▶ Designing in tolerance levels to cope with faults or deviations
▶ Limiting the effects of any adverse event
▶ Allowing for human error by designing in failsafe features

Q12 Describe the characteristics of:
(a) a pool fire
(b) a jet fire
(c) a flash fire or fire ball

Q12 Answer

A pool fire is a fire burning above a horizontal and stable pool of vaporizing hydrocarbon fuel.

A jet fire is a flame which is being fed by hydrocarbons continuously being released with significant momentum in a particular direction.

A flash fire or fireball is when a dense cloud of vapour is formed by the release of flammable gases or liquids and this meets a source of ignition.

Revision questions and suggested outline answers for Sub-element 1.4

Q1   Explain what the purpose of a safety case/safety report is.

Q1   Answer

The requirement of a safety case/safety report is to provide evidence and information that present a clear, comprehensive and defensible argument that a system is adequately safe to operate in a particular context.

Q2   An offshore safety case has THREE overriding principles which must be demonstrated within the document. Outline what those three principles are.

Q2   Answer

1   The management system is adequate to ensure compliance with statutory health and safety requirements.
2   Adequate arrangements have been made for audit and for audit reporting.
3   All hazards with the potential to cause a major accident have been identified, their risks evaluated, and measures have been, or will be, taken to control those risks.

Q3   Outline FIVE of the underlying principles or issues which will be covered in an offshore safety case.

Q3   Answer – five of the following:

▶   Factual information about the installation itself including the plant and systems used, its location and external environment.
▶   How the management system will apply appropriate levels of control during each phase of the installation's life cycle including the design, construction, commissioning, operation, decommissioning and dismantlement stages.
▶   All hazards with the potential to cause a major accident have been, or will be, identified, their risks evaluated and measures have been, or will be, taken to control those risks.
▶   How a systematic process has been, or will be, used to identify all reasonably foreseeable major accident hazards that are applicable to the installation.

▶ Show what criteria have been, or will be, adopted for major accident risk assessment.

▶ A description of what measures will be taken to manage major accident hazards.

▶ Show that effective rescue and recovery arrangements have been planned for to cope with major accidents.

▶ A description of how the principles of risk evaluation and risk management are being applied to the design to ensure that major accident risks will be controlled.

▶ Show how the management system addresses the additional risks associated with combined operations.

▶ When the installation is reaching the end of its working life, the safety case will have to be revised to deal with decommissioning or dismantlement operations.

▶ An explanation of how inherently safe design concepts were considered and applied.

Q4 **An onshore safety report is made up of FIVE main sections. Describe what each of those sections is and what issues it covers.**

Q4 Answer

Section 1 – Descriptive information:

▶ An overview of the facility and its activities
▶ Information about dangerous substances in use at the facility
▶ Information about the surrounding environment

Section 2 – Information on management measures to prevent major incidents:

▶ Major Accident Prevention Policy (MAPP)

Section 3 – Information on potential major incidents

Section 4 – Information on measures to prevent or mitigate the consequences of a major incident

Section 5 – Information on the emergency response measures of a major incident

▷ Onsite emergency plan
▷ Offsite emergency plan

Revision questions and suggested outline answers for Sub-element 2.1

Q1 **In relation to contract management, outline what the main responsibility of the client is.**

Q1 Answer

Clients are obliged to protect their own employees, as well as the employees of the contractors they are engaging, from risks to health and wellbeing. The client also has a responsibility towards the contractor and the contractor's own staff for hazards that may occur as a result of the client's own activities. The client is responsible for the workplace the contractor is working in.

Q2 **In relation to contract management, outline what the main responsibility of the contractor is.**

Q2 Answer

The contractor is responsible for the safe methods of working they employ. Contractors also need to give an assurance that the work they carry out will not impact on a facility's operational safety.

Q3 **When a client is going through the process of selecting a contractor there are a number of issues which can be considered to ensure the contractor is suitable. Give SIX of those issues.**

Q3 Answer – six of the following

▷ Is the contractor adequately insured?
▷ Will the contractor undertake a risk assessment of the proposed contracted work?
▷ Are the health and safety policies and practices of the contractor adequate?
▷ Is the contractor's recent health and safety performance reasonable?

230

▶ Is the contractor's health and safety training and supervision adequate?

▶ Does the contractor consult with their workforce?

▶ Does the contractor or their individual employees hold a 'passport' or other type of certification in health and safety training?

▶ Have any enforcement notices have been served on the contractor?

▶ Can the contractor offer any independent assessment of their competence?

▶ Has the contractor any references from previous clients?

▶ Are the contractor's qualifications and skills relevant?

▶ Is the level of competency of the workforce adequate?

▶ Is there appropriate certification for any equipment that the contractor might intend to use?

▶ Is the selection procedure for sub-contractors by the contractor adequate?

▶ Is the contractor a member of a relevant trade or professional body?

▶ Is the contractor's financial viability adequate?

▶ Is the contractor's safety method statement adequate?

Q4 When a site or piece of machinery is handed over to a contractor for them to undertake their work, a comprehensive system of safety checks should be implemented. Outline what these might include.

Q4 Answer

▶ Electrical isolations with lock offs as appropriate

▶ Mechanical isolations with lock offs as appropriate

▶ Physical barriers to restrict access to non-authorized personnel

▶ Pre-cleaning of the area

▶ Pre-cleaning of the equipment

Revision questions and suggested outline answers for Sub-element 2.2

Q1 **Explain** what process safety management is and what its main aim is.

Q1   Answer

Process safety management is the management of hazards associated with the processing of products. The main principle is to reduce the number of incidents involving the release of highly volatile or toxic substances or mitigate the severity of such incidents.

Q2 **The key provision of process safety management is process hazard analysis. Explain** what this is and what its main aim is.

Q2   Answer

Process hazard analysis is a careful review of what could go wrong and what safeguards must be implemented to prevent releases of hazardous substances.

Q3 **When considering plant layout, outline** the most important safety factors to be considered.

Q3   Answer

- Prevent or mitigate an escalation of events
- Ensure safety of personnel within on site buildings
- Control access of unauthorized personnel
- Facilitate access of emergency services

Q4 **Outline** what sort of information the Dow/Mond indices offer which might assist in plant layout design.

Q4   Answer

The Dow/Mond indices are used to evaluate process plant hazards and rank them against existing processes or projects in order to provide a comparative measure of the risk of fire and explosion. They do this by assigning them incident classifications and allow objective spacing distances to be considered during the development phase of a process or project.

Q5   Explain what is meant by the 'domino effect' when considering plant layout design.

Q5   Answer

A domino effect is the escalation of a single incident into multiple consequences and effects. These domino effect scenarios can be used to foresee these consequences, and the knowledge used to design in factors to counteract them.

Q6   With regard to control room design, outline the two main aspects that should be part of the design consideration.

Q6   Answer

▶ The ability of the control room to withstand a major hazardous event such as a fire, explosion or release of toxic gas or smoke.

▶ The efficient and appropriate layout of the control room and its equipment to ensure the effective operation and control of the plant under any circumstances including an emergency.

Q7   Explain what is meant by the phrase 'temporary refuge integrity'.

Q7   Answer

Temporary refuge integrity is the ability to protect the occupants following a hazardous event for a specific time period such that they will remain safe until they decide there is either a need to evacuate the installation or to recover the situation.

Q8   Outline what aspects management of change controls should include to ensure changes to process and equipment are implemented as safely as possible.

Q8   Answer

▶ Include expert personnel to review the proposed changes to ensure that they will not result in any operations exceeding established operating limits

▶ Ensure that any proposed changes are subject to a safety review using hazard analysis techniques

▶ Have in place arrangements for the control of relevant documents

233

▷ Ensure that any changes in the operating envelope such as temperatures, pressures, flow rates, etc., are communicated to the operators and documented

Revision questions and suggested outline answers for Sub-element 2.3

Q1 Describe the role and purpose of a permit-to-work system.

Q1 Answer

A permit-to-work is a detailed document which describes specific work at a specific site at a particular time which is to be carried out by authorized personnel. It also sets out any precautions and control measures which are necessary to complete the work safely.

Q2 Outline the key features of a permit-to-work system.

Q2 Answer

▷ The clear identification of who is responsible for specifying any necessary precautions
▷ The identification of personnel who may authorize particular jobs (including any limitations to their authority)
▷ The clear identification of any work classified as hazardous
▷ Clear identification of tasks, risk assessment, duration of permitted tasks, additional or simultaneous activity and control measures
▷ Training and instruction regarding the issue, use and closure of permits-to-work
▷ Monitoring and auditing to ensure that the system is working as planned

Q3 Outline the objectives of a permit-to-work system.

Q3 Answer

▷ It ensures that proper authorization of designated work has been granted.
▷ It ensures that those people who are conducting the work know the exact nature of the task including hazards, restrictions, time limitations, etc.

▷ It specifies the controls and precautions necessary to undertake the work safely.

▷ It ensures that those in charge of the location are aware that the work is being carried out.

▷ It provides both a system of continuous control and a record that appropriate precautions have been considered and applied by competent persons.

▷ It affords the ability to display, to those who need to know, exactly what work is ongoing.

▷ It provides a procedural means of suspending work when this is necessary.

▷ It provides an ability to control work which might interact or conflict with ongoing operations or other permit-to-work activities.

▷ It provides a procedural means of handing over the work when that work covers more than one shift.

▷ It provides a procedural means of handing back the area or plant which has been involved in the work.

Q4 Describe the type of work you would expect a permit-to-work system to cover.

Q4 Answer

▷ Special operations and non-routine work such as inspection, testing, dismantling, modification or adaptation of processes as well as repair work and non-routine maintenance

▷ Work done by two or more individuals or teams where activities need to be co-ordinated

▷ Work which will take longer to complete than one shift and the work and responsibility for it needs to be formally handed over

Q5 With regard to other work that may be ongoing at the same time as a permit-to-work is being issued, explain what considerations should be made.

Q5 Answer

When a permit-to-work is being considered for a job, it is important that the person issuing the permit is aware of any other activity (either planned or already under way) which may interact with the work to be done under the permit.

Q6 With regard to the issuing of permits-to-work to contractors, explain what considerations should be made.

Q6 Answer

▶ Ensuring that the contractor and his/her employees fully understand the permit-to-work systems and the arrangements within them which apply to the site where they are to carry out the work

▶ Ensuring that all personnel from the performing authority and other users are fully trained and aware of any specific arrangements in force to make the job safe at the location where they are to carry out the work

▶ Ensuring that the contractor understands fully the principles of the permit-to-work systems within the industry

Q7 Describe what the term 'lock out' or 'tag out' refers to and what its aim is.

Q7 Answer

The term 'lock out' or 'tag out' refers to a procedure which is aimed at safeguarding someone who is working on or near plant or machinery from that plant or machinery unexpectedly starting up or releasing energy of some kind.

This procedure involves an authorized person disconnecting the plant or machinery from its energy source. This person then locks or tags the isolator in order to prevent anyone from re-energizing the plant or machine.

Revision questions and suggested outline answers for Sub-element 2.4

Q1 Outline what is involved in a shift handover process.

Q1 Answer

The shift handover process is the effective transfer of information between the outgoing and incoming parties with no miscommunication or misunderstanding.

Q2 **The communication of information during shift handover is a two-way process. Describe this two-way process.**

Q2   Answer

The person who is providing the information needs to be sure that the person who is receiving it is in no doubt about the message being conveyed. It is a mutual interaction between the two parties – person(s) A (the sender of the information) and person(s) B (the person(s) receiving the information). Person(s) A needs to receive feedback from person(s) B confirming that the information has been received and correctly interpreted in order for the two-way communication to have been successful.

Q3 **Effective shift handover involves three specific periods of activity. Describe what these periods of activity involve.**

Q3   Answer

1   A period of time where the outgoing team prepares the information it will be conveying to the incoming team.

2   A period of time where both the outgoing and incoming team communicate with each other and exchange all relevant information.

3   A period of time where the incoming team cross-checks the information passed on to it as it takes on the responsibility for ongoing operations.

Q4 **Shift handover involves a number of important principles. Outline what these principles are.**

Q4   Answer

▶ To be treated as high priority

▶ Not to be rushed but to be allowed as much time and resources as necessary to ensure the accurate communication of information

▶ To be conducted using both verbal and written means of communication

▶ To be conducted face to face with both parties taking joint responsibility for the effective communication of necessary information

237

▶ To be conducted in an environment which is conducive to good communication without distractions

▶ To involve all shift personnel

Q5 Outline what kind of issues might be covered in shift handover involving plant and equipment.

Q5 Answer

▶ Work permits – the status of existing permits and the status of work in progress

▶ The updating of work permits

▶ Preparations for upcoming maintenance

▶ New personnel to the shift

▶ Any plant overrides – existing and planned

▶ Information about any abnormal events

▶ Any existing or planned shutdowns

▶ Any changes in plant parameters

▶ Any routine operations and existing parameters which may need to be carried out by personnel from the incoming shift

▶ Any breakdowns which may have occurred

▶ Any faults which have occurred with safety critical equipment

▶ Inhibits to the fire and gas and emergency shutdown systems

▶ Any completed work and equipment which has returned to service

Revision questions and suggested outline answers for Sub-element 2.5

Q1 Explain what is meant by the phrase 'asset integrity'.

Q1 Answer

Asset integrity is an asset's ability to function effectively and efficiently without creating undue hazards to any persons or the environment.

Q2 **Describe** what is meant by the phrase 'safety critical elements'.

Q2   Answer

Safety critical elements are those safety devices which are incorporated within an asset so that if a hazardous situation or incident does occur they are in place to deal with the consequences by quelling, controlling or mitigating the event or situation.

Q3 **Give FOUR examples of safety critical elements.**

Q3   Answer

▷ Blow out preventers
▷ Fire deluge systems
▷ Emergency shutdown valve
▷ Fire and gas detection systems

Q4 **Describe** what 'risk based management' is.

Q4   Answer

Risk based management is a technique used to assess the probability of various failure scenarios and applying an appropriate maintenance schedule in order to pre-empt these failures.

Q5 **Describe** what 'condition monitoring maintenance' is.

Q5   Answer

Condition monitoring maintenance involves taking readings of equipment in ways that allow those readings to give an insight into the condition of the plant or equipment without having to shut it down or dismantle it. For example, equipment which has rotating components can be monitored using vibration analysis.

Q6 **Give THREE examples of situations when a pre-start-up review should be conducted.**

Q6   Answer

▷ Before the initial start-up of a new installation
▷ When new chemicals or other hazardous materials are introduced into a process
▷ When existing facilities have had significant modifications or a maintenance shutdown

239

Q7 Explain what gas freeing and purging operations involve.

Q7 Answer

Gas freeing and purging operations involve filling spaces in plant or equipment that are currently filled with vapour, with an inert gas. This enriches the atmosphere in the vapour space with inert gas so that the atmosphere is taken to a level below its lower flammable limit.

Q8 Explain what inerting operations involve.

Q8 Answer

Inerting operations involve the partial or complete replacement of a flammable or explosive atmosphere in a contained environment with an inert gas. Inerting is typically used in storage tanks where a material may be above its flash point. Inert gases are also used in the transfer of flammable liquids under pressure, such as the transfer of hydrocarbons from seagoing oil tankers to land based facilities.

Q9 Explain what venting operations involve.

Q9 Answer

During any process operation, there will always be a need to vent the system if an overpressure problem occurs which will then activate a pressure relief valve. The venting system is made up of a series of pipes which are connected to each pressure relief valve. These pipes then channel the overpressure through to the vent stack where the overpressure is released to atmosphere in a safe and controlled manner (blown down). Alternatively, the overpressure can go to a flare where it is allowed to combust (burn) in a controlled manner.

Revision questions and suggested outline answers for Sub-element 2.6

Q1 Outline what the 'start-up and shutdown operating instructions and procedures' document should take into consideration.

Q1 Answer

▶ There should be no easier, more dangerous alternative procedures other than that specified in the operating procedure documentation.

240

▷ Operating procedures should include information about required personal protective equipment.

▷ Operating procedures should include an overview of the work to be done, any risks which the operator may be exposed to and a clear statement of any prerequisites.

▷ The document should be dated and an expiry date indicated where appropriate.

▷ The document should be clear as to which procedures apply to which situations.

▷ The document should be unambiguous.

▷ An appropriate method of coding each procedure should be used.

▷ Ensure that the most important information on the page is made the most prominent on the page

▷ Ensure that different sub-tasks are listed under separate headings to differentiate them.

▷ The document should be written in a language which is simple and familiar to the operators who will be performing the tasks.

▷ Ensure that the terminology used is consistent with that used on controls or panels.

▷ Ensure that any warnings, cautions or notes are put immediately before the instruction step to which they refer.

▷ Ensure that any shapes, symbols and colours which are used for graphics are consistent and conform to industry standards.

Q2 Explain what 'thermal shock' is and how its effects can be reduced.

Q2 Answer

Thermal shock is where a material is exposed to a sudden and significant change in temperature. This results in the material expanding at different rates within a limited area, causing a crack or failure. Thermal shock can be reduced by:

▷ The gradual introducing of steam or warm product from a lower temperature base

▷ Thoroughly warming up the systems prior to use

▷ Designing expansion loops into the system

▷ Using materials with greater thermal conductivity

241

▷ Reducing the coefficient of expansion of the materials
▷ Increasing the strength of the materials

Q3 **Explain how hydrates can be controlled within a process system.**

Q3   Answer

Hydrates can be controlled by introducing antifreeze (usually methanol or glycol) into the process system.

Q4 **Outline THREE methods of removing water from processed oil and gas.**

Q4   Answer – three of the following

▷ Where product is a mixture of oil and water and is allowed to stand for a while, water will naturally settle to the bottom as the specific gravity of water is greater than oil. This water can then be drained off.

▷ A centrifuge uses the principle of centrifugal force and the fact that water has a different specific gravity from oil; that difference allows this system to separate the two materials.

▷ Absorption removal is the process of using filters to absorb the moisture from the product as the product passes through the filter.

▷ Vacuum dehydration works on the principle that water boils at a lower temperature when it is at a lower pressure. Consequently, at 0.9 bar water will boil at around 52°C. Vacuum dehydration units reduce the pressure of the product within it. Air which has been dried and warmed is then passed over the product and the moisture is then transferred to the air from the product in the form of steam vapour.

▷ Air stripping is another form of vacuum dehydrator which works by mixing air, or nitrogen gas, with a stream of heated product within the air stripping unit. The gas then absorbs the moisture from the product and when the gas and product mixture is expanded, the gas separates from the product and takes the moisture with it.

▷ Heating the oil dry: some processes, because they run at elevated temperatures, are self-cleansing due to the fact that water naturally evaporates at these temperatures.

242

Q5 **Explain what 'commissioning' a process plant involves.**

Q5 Answer

Commissioning a process plant involves undertaking tests on the plant prior to it going into production in order to determine that it will function adequately and safely. The commissioning process also includes training the people who will operate the plant, as well as the control room staff who will oversee operations. Finally, it includes writing the operating procedure document.

Revision questions and suggested outline answers for Sub-element 3.1

Q1 **Describe the following words or phrases in relation to materials:**

(a) tension

(b) compression

(c) shear

Q1 Answer

Tension is a force which is acting in two opposite directions.

Compression is the opposite of tension in that the material is being compressed by pressure.

Shear is where two forces are being applied to a material but in opposite directions. The shear force acts as if it is trying to tear the material apart.

Q2 **Describe the following words or phrases in relation to materials:**

a) ductile

b) malleable

c) brittle

d) elasticity

Q2 Answer

A material which is ductile is one which can be subjected to tensile forces without it fracturing.

A material which is malleable is one which can have its shape changed without it cracking.

A material which is brittle is one which has no plastic deformation characteristics.

A material which has elasticity is one which can resume its former shape or dimension after a deforming force is applied and then released.

Q3   Describe the following words or phrases in relation to materials:

(a) creep
(b) stress
(c) stress corrosion cracking
(d) thermal shock
(e) brittle fracture

Q3   Answer

Creep is where a solid material is subject to long term exposure to high stress levels and gradually deforms in shape or dimension.

Stress is created when a load is applied to a material.

Stress corrosion cracking is where a material is subjected to both stress and corrosion. Stress Corrosion Cracking (SCC) can be highly chemical-specific with certain pairings of materials. These include:

▸ Brass paired with ammonia
▸ Stainless steel paired with chlorides
▸ High strength steel paired with hydrogen

Thermal shock is the stress introduced into a material as a result of a sudden and dramatic change in temperature.

An example of brittle fracture would be: if someone tried to bend an engineering file, which is made of a very hard but brittle metal, it would snap.

Q4   Explain what is meant by a 'safe operating envelope'.

Q4   Answer

A safe operating envelope is the parameters and conditions a plant must operate within to ensure it is not subjected to excessive stress which might introduce or encourage failure modes.

Q5 **Outline FOUR measures that can be applied when designing a plant to control stresses in material.**

▷ Fitting expansion loops and expansion bellows in steam pipes

▷ Ensuring the materials used are correct and capable of coping with the conditions they will be exposed to

▷ Ensuring the thicknesses of the materials used are adequate to cope with the stresses they will be exposed to

▷ Ensuring the strength of materials used is adequate

▷ Ensuring equipment is correctly supported

▷ Ensuring pipework and machinery are correct aligned

▷ Ensuring automatic shutdown trips are set to activate where parameters are exceeded

▷ Ensuring the control systems are adequate

▷ Ensuring operating procedures are always to hand

Q6 **Outline what stresses and conditions might cause the failure of the annular rim of a storage tank.**

The annular plate usually sits on a foundation of hardcore or a concrete ring wall, and is joined to the walls of the tank. Although the annular plate is not subject to high levels of stress, the joint of the plate to the walls is. This is because the weight of the product within the tank wants to push the walls outwards whilst at the same time push the annular plate downwards. This creates a high level of bending stress. The quality of the foundations will have a bearing on the downward deflection of the annular plate.

A further complicating factor to this stress is the fact that annular plates are prone to corrosion attacks both on the outer side where the tank shell sits on the annular rim and on the underside of the annular plate where trapped water may lie undetected. This corrosion, coupled with the prolonged stress, can lead to stress corrosion cracking and failure occurring without warning.

Revision questions and suggested outline answers for Sub-element 3.2

Q1   Give FOUR causes of weld failure.

Q1   Answer – four of the following

▶ Improper design of weld joint

▶ Poor selection of base materials and filler materials

▶ Inappropriate welding processes

▶ Residual stresses

▶ Ineffective or non-existent inspection procedures

▶ Welded components operating outside their safe parameters

Q2   Explain what porosity is in relation to welds and how it occurs.

Q2   Answer

Porosity is the presence of air bubbles within a weld. Porosity happens when the molten weld pool absorbs gases which become trapped and result in tiny bubbles once the weld solidifies. The main reason for this absorption is poor gas shielding during the welding process. Contamination from hydrogen can be attributed to moisture being present on the electrodes, the fluxes or the components being welded.

Q3   Describe the following non-destructive weld testing techniques:

(a) magnetic particle inspection

(b) dye penetration inspection

(c) ultrasonic flaw detection

(d) radiography

Q3   Answer

Magnetic particle inspection is where a magnetic field is introduced into the material to be tested. Ferrous iron particles coated in fluorescent dye are then applied and any defects will attract these particles to the location. These areas will be highlighted when an ultraviolet light is shone onto the area being tested.

Dye penetration inspection can detect surface-breaking defects in non-porous materials. The principle behind the process is that dye will penetrate into any surface-breaking defect and highlight it.

Ultrasonic flaw detection uses energy waves (high frequency vibrations – ultrasound) to read the exact structure of a component and thus detect any flaws or defects that may be present. The readings from the test procedure are displayed in a spiked graph form.

Radiography uses short-wavelength electromagnetic radiation to penetrate the material being inspected and highlight any defects. This electromagnetic radiation is emitted from one side of the component and detected and measured on the opposite side of the component. This allows an analysis of the composition of the material to be made. This is usually in the form of a film or negative (X-ray).

Revision questions and suggested outline answers for Sub-element 3.3

Q1 Identify the general groups of components of an emergency shutdown system.

Q1 Answer

▷ Various sensors to detect any fire or escape of gas or vapour
▷ Valves and trip relays to isolate sections of the process
▷ A system logic for processing any incoming signals
▷ An alarm system to warn operators and control room staff of a potential adverse occurrence

Q2 Give FIVE typical actions an emergency shutdown system might perform.

Q2 Answer – five of the following

▷ Shut down of part systems and equipment
▷ Isolate hydrocarbon inventories
▷ Isolate electrical equipment
▷ Stop hydrocarbon flow
▷ Depressurize/blow down

- ▶ Activate fire-fighting controls (water deluge, inert gas, foam system, water mist)
- ▶ Activate emergency ventilation control
- ▶ Close watertight doors and fire doors

Q3   Explain what the 'voting system' is within the response section of an emergency shutdown system.

Q3   Answer

An Emergency Shutdown (ESD) system is split and usually uses a triplicated microprocessor logic system. The system is based on an analysis of the information by voting on the inputs of signals received from fire and gas detectors positioned throughout the facility. Actions taken will depend on a voting system of input signals from the detectors. One vote out of three will raise an alarm which will be investigated. Two votes out of three votes will activate the emergency shutdown system.

Q4   Explain what 'safety integrity levels' are.

Q4   Answer

The components of a safety instrumented system need to have a level of integrity or dependency which is in line with the consequences of failure. The greater the consequence, the higher level of integrity or dependency needed.

There are four discrete integrity levels, SIL1 to SIL4. The higher the Safety Integrity Level, the higher the associated safety level requirement is, and this needs to be coupled with a lowering of the probability that the safety instrumented system will fail.

Q5   Explain under what circumstances an emergency shutdown system might need to be bypassed.

Q5   Answer

All emergency shutdown systems and fire and gas systems need to be tested, inspected and/or maintained on a regular basis to ensure they are functioning as required. These testing and/or maintenance procedures involve the temporary bypassing of safety system interlocks. These are the devices within the logic system which

248

activate the alarm as well as sending a signal to the actuator on the emergency shutdown component itself.

Q6 **When an emergency shutdown system is bypassed this should be recorded in an inhibit log. Identify FIVE pieces of information relating to the bypass that should be recorded in the log.**

Q6   Answer

▶ The safety function being inhibited
▶ The time and date the inhibit was applied
▶ A cross reference to the relevant permit-to-work or protective systems isolation certificate where applicable
▶ The time and date of each reassessment
▶ The time and date the inhibit was removed

Q7 **Describe what the term 'blow down' refers to.**

Q7   Answer

The term 'blow down' refers to the action of venting gas or relieving pressure from a process or production system.

Q8 **Describe what an 'open drain system' is and its function.**

Q8   Answer

A drainage system which collects fluids that spill onto the ground is called an open drain. This system will consist of collecting funnels or trays known as 'tundishes' which are strategically positioned so as to channel any liquid they collect through a series of drain pipes to an open drain header before being routed to a slop tank ready to be dealt with appropriately. There will be separate drainage systems to deal with 'safe areas' and 'hazardous areas'.

Q9 **Describe what a 'closed drain system' is and its function.**

Q9   Answer

A drainage system that is required to be connected to a pressure vessel in order to drain off fluids is called a closed drain system. Whenever a vessel is drained of liquid under pressure into a pressure or closed drain system, it must be assumed that the liquid contains a certain amount of dissolved gases. Furthermore, the flow of liquid from the pressurized vessel will be followed by a certain amount

249

of gas (known as gas blow by). This gas, in both its forms, can represent a hazard if it is not dealt with appropriately, e.g. using a blow down facility so that gas is vented away to a flare.

Q10 Describe what an 'interceptor' is and its function.

Q10 Answer

Interceptors are a means of collecting contaminated water before it is discharged to a foul drain or surface drain. Typically, interceptors have three separate chambers, with the divisions between chambers extending down to the bottom, and low level pipes connecting the chambers. This is so that when the contaminated water enters the first chamber it can separate (oil will naturally float on top of water) and be extracted. The water is then directed to the second and third chambers via the low level pipe where any residual oil is also allowed to separate and be extracted. Finally, the water from chamber three is channelled into either a foul or surface drain, whichever is appropriate.

Revision questions and suggested outline answers for Sub-element 3.4

Q1 Explain why the maximum volume of product stored in a storage tank can vary.

Q1 Answer

The density of oil varies from approximately 0.8 tonnes per m³ to 1.3 tonnes per m³ depending on its type and grade, and the prescribed volume within a storage tank will have to be adjusted accordingly.

Q2 Outline SIX management practices which can be regarded as good with regard to maintaining the integrity of storage tanks.

Q2 Answer – six of the following

▶ Tanks containing hazardous substances should be recorded in the plant register.

▶ Operators should maintain tank data files.

- Compatibility assessments should be undertaken where tanks are used for multi-product service.
- Tanks should be subject to formal periodic maintenance and inspection.
- Inspection and maintenance of tanks should only be carried out by qualified persons.
- Appropriate inspection techniques should be utilized, depending on deterioration mechanisms.
- Inspection reports and checklists should be of high quality.
- Recommendations from inspection reports should be actioned promptly.
- Assessment for fitness for service should be carried out following significant changes to process or operating conditions.
- Operators and competent persons should have knowledge of, and adopt the recommendations given in, the relevant guides, codes and standards.
- Tank examination schemes should include both internal and external inspections.

Q3   With regard to storage tanks, explain what the process of 'corrosion' is and where it can take place.

Q3   Answer

Corrosion is the gradual deterioration of a substance by way of a chemical reaction with its immediate environment. Most storage tanks are made from carbon steel, which makes corrosion a primary cause of deterioration. Corrosion can occur both internally and externally.

Q4   With regard to storage tanks, explain what the process of 'erosion' is and where it can take place.

Q4   Answer

Erosion is the process of material being worn away by the constant movement of product flowing over the surface. Areas such as filling and discharge points which experience large amounts of product flow are the most vulnerable points.

Q5   With regard to thermoplastic storage tanks, explain what the process of 'creep' is.

Q5   Answer

Creep is where the tank material stretches over a period of time when it is under stress from the weight of the product being stored. However, this condition only affects non-metallic thermoplastic tanks. Increases in temperature tend to compound the problem, with high-density polyethylene (HDPE) being particularly vulnerable and losing its strength as the temperature increases.

Q6   With regard to storage tanks, explain what the process of 'fatigue' is and where it can take place.

Q6   Answer

Pressure vessel storage tanks are more prone to fatigue defects than tank storage vessels. This is because the membrane stresses in pressure vessels are much greater.

For tank storage vessels which are subject to a cycle of frequent filling and emptying, certain parts of the tank's construction can be vulnerable, particularly welded joints.

Q7   With regard to storage tanks, give TWO examples of how 'settlement which is non-uniform' can happen.

Q7   Answer – two of the following

▶ Subsidence can be caused by constant weight on weak or compressible terrain. The weakness or compressibility may not be uniform, which can lead to distortions in the tank structure.

▶ Frost heave, where the ground is subject to frequent freezing and thawing, can cause non-uniform settlement.

▶ Ground movement caused by high tides in areas close to the sea can cause non-uniform settlement.

Q8   Explain under what circumstances it is possible for a storage tank to be floated off its foundations.

Q8   Answer

Where rainwater is allowed to build up inside a bunded area with a tank in it which has not got a lot of product within it and which has

252

not been anchored down, it is possible for the tank to float off its foundations.

Q9   Explain what 'landing the roof' of a storage tank is and the hazards associated with it.

Q9   Answer

Landing a roof is where the liquid in a tank falls far enough for the legs on the underside of the floating roof to land on the base of the tank. The void between the liquid and the roof will grow which will allow a build-up of vapour to occur. This has the potential to cause a fire and/or an explosion.

Q10 With regard to external roof storage tanks, give TWO reasons for a floating roof of a storage tank to sink.

Q10 Answer – two of the following

▶ Build-up of rain or snow
▶ Use of access ladders on one side of the roof
▶ Earthquake activity may have a destabilizing effect on the roof

Q11 Describe what safety features can be incorporated into a floating roof storage tank which will address the issue of a rim seal failure.

Q11 Answer

Most systems have a double seal so that there is a back-up in place should the primary seal fail. Furthermore, there is usually a detection system built onto the roof so that if a leak does occur, it can be detected and dealt with. The detection system may well incorporate an automatic rim seal fire-fighting system as well.

Q12 With regard to fixed roof storage tanks, explain how 'over-pressurization' can occur, its effects and the control measures required to prevent it happening.

Q12 Answer

When liquid is pumped into a storage tank the atmosphere in the void needs to be vented out. If this does not happen the pressure will become unsustainable and the tank may rupture.

Other factors which can cause increased pressure within the tank include:

▶ Storing a volatile product will cause gases to evolve and increase pressure.

▶ Warm weather or direct sun on the tank will warm up the product and make it expand, thus increasing pressure.

Normally, tanks have a Pressure and Vacuum (P&V) relief valve fitted on or near the top of the tank. This allows vapour or gas to escape and ensures the integrity of the tank is retained.

Another method of controlling changes in pressure is to employ an expansion or pressure exchange system. This is where the expelled vapour or gas is directed to another tank rather than to atmosphere.

Q13 With regard to fixed roof storage tanks, explain how 'vacuum' can occur, its effects and the control measures required to prevent it happening.

Q13 Answer

When a tank is emptied the volume of liquid reduces and the void above the liquid needs to have air vented in so that a vacuum is not created. If the air inlet valve is faulty, and a vacuum is created, this may well cause the tank to collapse.

Cold weather will cool the product down and cause it to contract, thus decreasing pressure.

Normally, tanks have a Pressure and Vacuum (P&V) relief valve fitted on or near the top of the tank. This allows air to enter the tank and ensures the integrity of the tank is retained.

Another method of controlling changes in pressure is to employ an expansion or pressure exchange system. Where a vacuum has to be dealt with, the system draws in an inert gas from an external source.

Q14 Explain what bunding of storage tanks is and its purpose.

Q14 Answer

Bunds are designed to contain any spillages from storage tanks and stop them escaping.

Q15 With regard to the transfer of material between road/ rail tankers and storage tanks, give FIVE safety measures which should be incorporated into the procedure.

▶ Secure the road/rail tanker – the brakes should be applied and the engine turned off.
▶ Check that hoses are suitable for the product being discharged and the operating pressures.
▶ Ensure that connections of pipes and hoses are secure.
▶ Position drip trays beneath all connections.
▶ Closely monitor all connections during transfer operations.
▶ Ensure that a good bonding connection has been established between the tanker and loading/unloading equipment.
▶ Prior to transfer operations, make ready fire control measures.
▶ Make note of emergency escape routes before transfer commences.
▶ Monitor the flow rates and fill capacities.
▶ Have adequate venting arrangements in place to ensure vapour is dispersed properly and safely.
▶ Have venting arrangements to the recipient tank in place as well as similar arrangements for the donor tank.

Q16 Describe the characteristics of a 'jet fire'.

Q16 Answer

A jet fire is where a fire emanates from a source of fuel which is being continuously released in a particular direction with significant force.

Q17 Describe the characteristics of a 'pool fire'.

Q17 Answer

A pool fire is a fire which burns above a horizontal pool of vaporizing hydrocarbon fuel.

Q18 Describe how a Boiling Liquid Expanding Vapour Explosion (BLEVE) might occur.

Q18 Answer

Any vessel which is partly filled with pressurized hydrocarbon liquid will have a certain amount of space above it filled with vapour. If the vessel is subjected to a fire, the pressure in the tank will increase due to the liquid going above its boiling point and turning into a vapour.

The pressure relief valve will allow the overpressure to be vented to atmosphere in the first instance but this will reduce the amount of liquid in the tank still further and the potential for the flame to engage with a section of the tank containing vapour and not liquid will increase. If this happens, the tank wall will weaken at this point as the heat transfer to vapour is much less efficient than to a liquid.

The result is likely to be a sudden and catastrophic failure of the vessel, with a discharge of vapour followed by an explosion when it reaches the flames. This is a boiling liquid expanding vapour explosion.

Q19 Describe what a Confined Vapour Cloud Explosion (CVCE) is.

Q19 Answer

A confined vapour cloud explosion is an explosion following a leak of vapour which occurs in a confined space, such as a building or a tank.

Q20 Describe what an Unconfined Vapour Cloud Explosion (UVCE) is.

Q20 Answer

An unconfined vapour cloud explosion is an explosion following a leak of vapour which occurs in an unconfined space (outdoors).

Q21 Outline TWO examples of how pipelines can be damaged.

Q21 Answer – two of the following

Pipelines can be damaged by trailing anchors under the sea and trawling activities. Deliberate damage can also come from arson or terrorist attacks, or from illegal tapping.

Q22 Outline the FIVE stages of decommissioning.

Q22 Answer

1   All options for the physical dismantlement and/or removal of the installation are assessed and the best option is selected.

256

This is then put through a detailed planning process involving all engineering, safety and environmental aspects.

2   Production/extraction is stopped and all the wells are plugged.

3   The installation is dismantled and/or removed in line with the agreed plan of action.

4   Those parts of the installation that have been removed are disposed of or recycled.

5   A seabed survey is carried out to ensure nothing untoward has happened to the original location of the installation. If any part of the installation remains, ongoing monitoring will be implemented.

Q23 Outline the measures which should be taken to address potential conflicts when simultaneous operations are undertaken.

Q23 Answer

The parties involved in simultaneous operations should meet to discuss the extent of their work and their requirements. The intention of the meeting is for each party to be able to draw up a work-specific dossier which covers how their activities will be conducted whilst taking into account all the other parties' activities and avoiding any conflict. The meeting should resolve the following issues:

▶   The extent of the responsibilities of each party

▶   Nomination of a responsible person for each party

▶   Ensuring responsible persons know of each other and have contact arrangements in place

▶   Identifying the requirements of each party

▶   Identifying the time frame of the actual work activity of each party

Revision questions and suggested outline answers for Sub-element 3.5

Q1 Give TWO examples of potential threats from a lightning strike.

Q1 Answer

▶   Sparks which can cause a fire or explosion

▶   Power surges to electrical equipment, particularly monitoring and safety devices, which can render them inoperable.

Q2 Give the THREE elements which make up the fire triangle.

Q2 Answer

▶ Heat or a source of ignition
▶ Fuel
▶ Oxygen

Q3 Explain what 'thermal radiation' is.

Q3 Answer

This is the transfer of heat from one source to another. This can be a structure or a person.

Q4 Describe how electrostatic charges can be generated within hydrocarbon liquid product.

Q4 Answer

Whenever a liquid moves against a solid object, such as the inside of a pipe, it generates a static electrical charge. The most common cause of static electricity build-up is where there is a flow (transfer) or movement (mixing process) of liquid within a process.

Q5 Outline THREE factors which can influence the build-up of static charge within hydrocarbon liquid product.

Q5 Answer – three of the following

▶ The conductivity of the liquid
▶ The amount of turbulence in the liquid
▶ The amount of surface area contact between the liquid and other surfaces
▶ The velocity of the liquid
▶ The presence of impurities in the liquid
▶ The atmospheric conditions

Q6 Describe ONE control measure that can be implemented to reduce static charge in piping systems.

Q6 Answer

Keep the rate and velocity of the liquid low.

258

Q7 **Describe TWO control measures that can be implemented to reduce static charge when filling operations take place.**

Q7 Answer – two of the following

▷ Ensure that filling operations do not involve the free-fall of liquids.

▷ Lower the velocity of the liquid being filled.

▷ Ensure that fill pipes touch the bottom of the container being filled.

▷ Tanks which have been filled with products that have a low conductivity should be given time to relax before the process continues.

▷ Tanks which have been filled with product should not have any ullage for a set period of time. Nor should any dipping of the product take place, again for a set period of time.

Q8 **Explain why filtration of hydrocarbon liquid product can create high levels of electrostatic charge.**

Q8 Answer

Filters have large surface areas which can generate as much as 200 times the amount of electrostatic charge compared with a piping system without filtration.

Q9 **Explain how bonding and grounding of equipment can help to reduce the potential for static charges to be generated.**

Q9 Answer

A bonding system is where all the various pieces of equipment within a process system are connected together. This ensures that they all have the same electrical potential, which means there is no possibility of a discharge of electricity, by way of a spark, from one piece of equipment to another.

Grounding is where pieces of equipment are connected to an earthing point. This ensures any electrical charge in the equipment is given the means to constantly flow to earth, thus ensuring there is no potentially dangerous build-up of charge which could lead to a sudden discharge of electricity, by way of a spark.

259

Q10 Give FIVE examples of sources of ignition.

Q10 Answer – five of the following

▶ Smoking and smoking material
▶ Vehicles
▶ Hot work such as welding, grinding, burning, etc.
▶ Electrical equipment
▶ Machinery such as generators, compressors, etc.
▶ Hot surfaces such as those heated by process or by local weather (hot deserts)
▶ Heated process equipment such as dryers and furnaces
▶ Flames such as pilot lights
▶ Space heating equipment
▶ Sparks from lights and switches
▶ Impact sparks
▶ Stray current from electrical equipment
▶ Electrostatic discharge sparks
▶ Electromagnetic radiation
▶ Lightning

Q11 Explain what 'zoning' of hazardous areas is and why it is implemented.

Q11 Answer

A place where an explosive atmosphere may occur on a basis frequent enough to be regarded as requiring special precautions to reduce the risk of a fire or explosion to an acceptable level is called a 'hazardous place'. Determining which areas are hazardous places, and to what extent, is called a 'hazardous area classification study'. Hazardous areas are classified into zones based on an assessment of two factors:

1 The frequency of the occurrence of an explosive gas atmosphere
2 The duration of an explosive gas atmosphere

These two factors in combination then facilitate the decision-making process which will determine which zone will apply to the area under consideration.

260

Q12 Explain why electrical tools need to be categorized as suitable for use in specifically zoned hazardous areas.

Q12 Answer

Apparatus, tools and equipment are categorized in accordance with their ability to meet the standards required when used within each zone.

Revision questions and suggested outline answers for Sub-element 3.6

Q1 There are two main types of boiler. Describe how both work and how they differ.

Q1 Answer

A fire tube boiler is one that generates hot gases which then pass through a number of tubes before being expelled out of the flue. These tubes run through a sealed and insulated container of water and the heat from the gases is transferred by thermal conductivity to the water, which then turns to steam. The steam from the boiler then exits through a tube at the top of the container.

Water tube boilers have basically the opposite configuration of fire tube boilers. In a water tube boiler, a number of tubes run through the furnace part of the boiler. This heats the water inside the tubes, which consequently turns to steam.

Q2 Outline the hazards associated with pilot lights in boilers and furnaces, and the safety measures which will address these hazards.

Q2 Answer

Pilot lights provide a source of ignition to the main boiler when it requires to be fired up. In a situation where the pilot light fails to light or goes out, gas will continue to enter the chamber, causing a build-up of flammable gas which, if ignited, could cause an explosion.

In order to counteract this scenario, a sensory device known as a thermocouple is located in close proximity to the pilot light in order to detect if there is heat coming from the pilot light. If no heat is

261

detected, the device will activate a relay which will close the main gas valve.

Q3 **Explain what 'over-firing' is with regard to boilers and furnaces.**

Over-firing is basically allowing the heat flux to increase to a level beyond its upper 'maximum continuous rating', which is set by the manufacturer of the boiler. This can then have an impact on the furnace walls and the surface temperature of the refractory. It can also result in a substantial increase in tube and membrane operating temperatures, which can lead to a degradation of tube metallurgy and strength.

Q4 **Explain what 'flame impingement' is with regard to boilers and furnaces, and how it can be controlled.**

Flame impingement is where the flame produced by the burners comes into contact with the surface being heated. If this occurs there tends to be a gradual build-up of carbon on the inside of the tube at the point where the flame is in contact. If this process is allowed to continue, it can lead to the tube eventually becoming blocked, resulting in the potential rupture of the tube.

Q5 **Explain what 'firebox overpressure' is with regard to boilers and furnaces.**

Firebox overpressure typically occurs after a flameout, which is where the burner flame is extinguished for some reason. The fumes, gases or vapour from the fuel will begin to build up inside the combustion chamber, which will invariably be hot. This will make them highly volatile and when they reach their Lower Explosive Limit (LEL) and make contact with a source of ignition, this will cause overpressure (explosion).

Q6 **Explain** **what 'low tube flow' is with regard to boilers and furnaces, and how it can be controlled.**

Q6 Answer

Normal circulation within water or steam pipes heated by a boiler is generated by the difference in density between cooled water and hot water/steam. Any adverse conditions within the flow system will affect this flow rate and create a situation where flow rates are eventually reduced. When this becomes acute, it is known as 'low tube flow'.

Q7 **Explain** **what 'tube material temperature' is with regard to boilers and furnaces, and how it can be controlled.**

Q7 Answer

This is where localized overheating occurs which can lead to potential component failure. It is generally due to poor water quality where suspended material tends to congregate in the bottom of the boiler and cause scale to develop. It can also be caused by ingress of product (e.g. oil) into the condensate return system.

Controls include regular testing of water and blowing down on a regular basis, as well as checking for leaks on the water/steam circulation system.

Q8 **Explain** **what 'Total Dissolved Solids' (TDS) are with regard to boilers and furnaces and how they can be controlled.**

Q8 Answer

Total Dissolved Solids (TDS) are substances such as minerals, salts and metals which are held in a suspended form within water. If these solids are of a sufficient concentration within the water used in a boiler system, they can attach themselves to the inside of boilers and, over time, build up to form scale.

The first control is to maintain the solids below a certain limit. This is done by testing the water with a TDS meter or conductivity meter.

If the TDS concentration increases, the likelihood that the dissolved solids will precipitate out of the water and form scale also increases. If this happens, it is necessary to drain some of the water from the system, called boiler blow down, in order to remove some of those dissolved solids and keep the TDS concentration below the level where they will precipitate.

Revision questions and suggested outline answers for Sub-element 4.1

Q1 **Give FIVE examples of components within a fire and gas detection system.**

Q1   Answer – five of the following

▶ Spot fire detector

▶ Camera based flame detector

▶ Ultraviolet flame detector

▶ Combined ultraviolet and infrared flame detector

▶ Point heat detector

▶ Linear heat detector

▶ Infrared absorption combustible gas detector

▶ Point infrared gas detector

▶ Open path infrared gas detector

▶ Ionization point smoke detector

▶ Optical point smoke detector

▶ Optical beam smoke detector

▶ Aspiration smoke detector

▶ Leak detection systems

Q2 **Explain the following**

(a) fire compartments

(b) detection zones

(c) alarm zones

Q2   Answer

Buildings are generally divided into sections, known as fire compartments, which use fire resistant walls, ceilings and floors to enclose them in order to limit the spread of any fire that may break out within any one of these sections.

Detection zones are essentially a convenient way of dividing up a building, area or plant into manageable sections so as to assist in quickly locating the position of any fire or leak.

264

Alarm zones are where it is necessary to operate alarm devices differently in various parts of the building. These should be divided into alarm zones so that all of the alarm devices in one alarm zone operate in the same way.

Q3 **Explain** what is meant by the phrase 'passive fire protection'.

Q3   Answer

'Passive fire protection' covers the materials, products and design measures built into a building or structure in order that any fire which may start in the building or structure is restricted in its growth and spread. This is achieved by controlling the flammability of the structure, including walls, ceilings, floors, doors, etc., as well as by protecting structural steel members from severe heat which might compromise their integrity.

Q4 **Fire walls are coded to indicate their ability to withstand fire. Explain how this code is made up and what it represents.**

Q4   Answer

Walls and divisions which have been manufactured as passive fire protection are coded in relation to the type of fire they are designed to withstand and its duration. The code is made up of a letter (either H or A) followed by a number (0, 30, 60 or 120). The letters indicate the type of fire they are designed to withstand. H is for hydrocarbon fire and A is for cellulosic fire. The number which follows the letter indicates the length of time in minutes that the wall or division has been designed for to hold back the fire.

Q5 **Explain** what is meant by the phrase 'active fire protection system'.

Q5   Answer

Active based fire protection systems are those fire protection systems which are primed and ready to be activated on a given signal.

Q6 Foam, when used as a means for fire-fighting, has a number of controlling effects. Outline THREE effects.

Q6 Answer – three of the following

▶ The foam cover separates the combustion zone from the surrounding atmosphere and, as such, prevents oxygen feeding the fire.

▶ The foam cover stops the evaporation of flammable vapours from the burning material. This means flammable vapours, which would normally be encouraged by the heat of the fire and which feed the fire, are stopped at source.

▶ When foam is applied to a liquid fire, water is gradually discharged by the foam, which adds a cooling effect to the flame.

▶ High- or medium-expansion foam used to flood areas will suppress the release of flammable vapours/gases from combining with air which is necessary for the combustion process to take place.

▶ Foam has a low thermal conductivity level. Consequently, when foam covers any flammable liquid which is not burning, it tends to insulate it from thermal radiation and ignition.

▶ Foam which is used on non-polar hydrocarbons (the majority of hydrocarbons) produces a thin aqueous film which helps the foam flow, as well as assisting in the extinguishing process and inhibiting re-ignition.

Q7 Describe how a chemical fire extinguishing system using dry powder works.

Q7 Answer

The chemical reacts with the fire and extinguishes it.
Monoammonium phosphate based chemicals melt at a relatively low temperature, thus blanketing the burning surfaces and preventing re-ignition.

Q8 Explain how a water mist fire extinguishing system works.

Q8 Answer

When a water mist system is activated, water is discharged in micron sized droplets. These droplets rapidly convert the energy in the fire to steam, which starves the fire of oxygen.

266

Q9 Give an example of a fire protection system for storage tanks.

Q9   Answer

A linear heat detection system is placed at the rim seal area of the floating roof, and this is connected to a number of foam based fire extinguishing modules positioned around the perimeter of the roof. When the linear heat detector registers a fire, as well as raising the alarm it will also activate the fire extinguisher module nearest to the fire. This will then flood the immediate area with foam and extinguish the fire.

Revision questions and suggested outline answers for Sub-element 4.2

Q1 Outline what the general contents of an emergency plan include.

Q1   Answer

- Responsibilities and authority of those overseeing an emergency
- Types of events planned for and extent of responses planned
- Alarm systems and responses to alarms
- Arrangements for triggering any off-site emergency plan
- Training and instructions
- Off-site communication measures

Q2 Explain what a 'fire and explosion strategy' is.

Q2   Answer

This is a combination of measures taken to reduce the risk of fire and explosion happening.

Q3 Outline SIX measures which will be considered within a fire and explosion strategy.

Q3   Answer – six of the following

- Buildings which are occupied should have an assessment made of the risks and hazards they might be vulnerable to if a major incident occurred.
- Escape routes should be clearly marked.
- All escape routes, where appropriate, should be protected.

267

▷ Escape routes should be of a size that is adequate to accommodate all personnel.

▷ Where appropriate, the installation should be compartmentalized.

▷ Where appropriate, blow out, or explosion panels, should be strategically positioned.

▷ Where appropriate, escape routes should have heat activated deluge/sprinklers within them.

▷ Each area of the installation should have more than one escape route.

▷ Escape routes should be protected against the effects of fire and explosion.

▷ There should be internal access to the helideck from any temporary refuge facility.

▷ There should be a policy of ensuring the number of overrides and inhibits applied to the Emergency Shutdown (ESD) system and the Fire and Gas (F&G) system is kept to a minimum.

▷ At the design stage of an Emergency Shutdown (ESD) system, failsafe and fireproof ball valves should be incorporated.

▷ At the design stage of a process system the amount of flanged pipework should be minimized.

▷ At the design stage the inventory of hazardous substances should be reduced to a minimum.

▷ Emergency Shutdown Valves (ESDVs) should be enclosed with fireproof casing.

▷ Water deluge operating skids should be situated away from the area they are protecting.

▷ Access doors to accommodation areas should have automatic door closers fitted.

▷ All enclosures which house rotating equipment and electric drives should have what is known as 'very early smoke detection'.

▷ Measures should be taken to ensure that the mechanical and natural ventilation to production areas is sufficient to assist in dispersing any gas leak.

▷ The accommodation and control rooms should be segregated and distanced away from production processes.

▷ The control rooms and emergency command and control centres should be segregated using blast and fire walls.

- Subsea Isolation Valves (SSIVs) should be fitted in sea lines and wells.
- High Integrity Pipeline Protection Systems (HIPPSs) should be fitted where appropriate.
- External fire protection should be fitted to the accommodation rooms and temporary refuge.
- The Temporary Refuge (TR) should be airtight and always under positive atmospheric pressure.
- There should be a separate Emergency Command and Control (ECC) centre in the Temporary Refuge (TR) when the control room is not situated within the TR.

Q4 Explain the difference between an alarm used on onshore facilities and an alarm used on offshore facilities.

Q4 Answer

An onshore alarm signal is conveyed by means of a warning siren which is loud enough for everybody on the site, and in the surrounding area, to hear.

Offshore, there are two types of alarm used. The first type of alarm is the 'general platform' alarm, which is a general alarm calling all personnel on board to go to their allocated muster station. The general platform alarm is an intermittent signal of a constant frequency. The second type of alarm is the 'prepare to abandon platform' alarm, which is sounded to inform personnel of the imminent evacuation of the platform. The prepare to abandon platform alarm is a continuous signal of variable frequency.

Q5 Explain what is meant by the phrase 'medical triage'.

Q5 Answer

This is a process of prioritizing casualties according to their medical needs.

Q6 Outline SIX factors which will enhance an escape and evacuation route from an offshore facility.

Q6 Answer – six of the following

- Escape routes should be clearly marked using high visibility signage along their entire route.

▷ Escape routes should be well lit and include emergency lighting in case power is lost.

▷ Escape routes should be protected, where possible, by fire walls or by deluge fire protection systems.

▷ Escape routes should be congestion free (have clear access and egress) and be adequate in size to accommodate all personnel.

▷ Escape routes should have heat activated deluge/sprinklers within them.

▷ Each area of the installation should have more than one escape route.

▷ Where appropriate, the installation should be compartmentalized (have fire walls between compartments).

▷ Where appropriate, blow out, or explosion panels, should be strategically positioned within the installation to alleviate any overpressure.

Q7 Give THREE examples of emergency evacuation from an offshore facility.

Q7   Answer – three of the following

▷ Lifeboat
▷ Life raft
▷ Knotted rope
▷ Sea ladder
▷ Scramble net
▷ Skyscape
▷ Helicopter
▷ Standby vessel

Q8 Describe what the underlying principle is with regard to training and drills for emergency response.

Q8   Answer

Training and drills, in relation to emergency response, are about ensuring everybody is in a state of preparedness and knows exactly what to do and what will be expected of them should an emergency arise.

Q9 Explain the purpose of undertaking drills in relation to emergencies.

Q9 Answer

Drills serve the purpose of training personnel in the practical application of their roles and responsibilities in an emergency.

Q10 Explain why it is important that the management of facilities liaise with emergency services.

Q10 Answer

In order to ensure any emergency is responded to effectively, it is essential that good channels of communication are established between the management of a facility and the emergency services.

Q11 Outline SIX pieces of information which will be required by the emergency services in the event of an emergency.

Q11 Answer – six of the following

- Contact point for the fire/police/ambulance liaison officer
- Contact point for fire/police/ambulance incident commander
- Rendezvous point for emergency services
- Strategic response group member (planning co-ordinator or as otherwise delegated) with primary responsibility for managing the emergency at the site
- Likely cause and effect of the emergency
- Likely casualty status including potentials, how serious and their current location
- Roll call results
- Map of the site including floor plans, entry and exit points
- Evacuation location
- Outline of the local environment and surrounding risks (possibility of secondary incidents/contamination)
- Utility shut-off points
- Availability of CCTV
- Press liaison details
- Welfare arrangements
- Traffic control points and likely impacts on the surrounding area

271

Revision questions and suggested outline answers for Sub-element 5.1

Q1 Describe what an 'exclusion zone' around an offshore facility is.

The immediate area around an installation is regarded as a major risk area with potentially severe consequences from any incident that happens within it. Because of this, it is designated as an exclusion zone with restrictions on which vessels are allowed to enter the zone.

The zone extends for 500 metres around the installation and is constantly radar monitored, as well as being patrolled by the platform's Standby Vessel (SBV) or its emergency response and rescue vessel, either of which is in close communication with the platform's Central Control Room (CCR). Any vessel wishing to enter the exclusion zone must seek permission from the central control room and the patrol vessel before doing so. If a vessel tries to enter the zone without permission, it will be warned off by the patrol vessel.

Q2 Outline THREE hazards associated with vessels in relation to offshore facilities.

- Breakdown, loss of power or loss of steering.
- Anchoring over pipelines, wells and submerged cables.
- Explosion during loading/unloading operations
- Pollution
- Striking the installation
- Man overboard

Q3 Explain what 'single buoy mooring' is and its purpose.

There are many remotely situated sea based well heads throughout the world which only have a buoy connected to them. These are known as Single Buoy Moorings (SBMs) or Single Point Moorings (SPMs). These buoys act as a mooring point for tankers and have within them a product transfer system which facilitates the transfer of hydrocarbon product to the tanker.

Q4 **The loading and unloading of vessels at marine terminals can be a hazardous event. Outline SIX procedures or controls which will minimize the risks associated with these operations.**

Q4   Answer – six of the following

- The vessel must be securely moored with sufficient mooring scope to ensure it does not range along or away from the berth.
- The mooring scope must also take into account tidal rise and fall, river currents and the possible effects of passing ships.
- Ensuring that hoses are suitable for the product being discharged and the operating pressures they will be subjected to.
- Ensuring that the connections of pipes and hoses to be used in the transfer operation are secure.
- Positioning drip trays beneath all connections and ensuring there is close monitoring of connections during transfer operations.
- Deploying fire wires on the vessel to give tugs a means of moving the vessel away from its berth quickly if an emergency arises.
- Reducing the risk of static electrical charges occurring on board ship by ensuring all metal objects are bonded to the ship.
- To protect against the risk of possible differences in electrical potential between the ship and the berth, there should be a means of electrical isolation at the ship/shore interface.
- Fire control measures, such as fire-fighting equipment, should be made ready before transfer commences.
- A discharge plan which ensures the vessel is not subject to undue internal stresses as the cargo is discharged.
- The control room should monitor flow rates and quantities.
- All doors and windows on board the vessel and in buildings at the terminal to be closed.
- Adequate venting arrangements should be in place to ensure vapour is dispersed properly and safely.
- Venting arrangements should be made for both the recipient tank and the donor tank.

▷ Once discharge commences, the vessel must be kept within the operating envelope (limits) of the oil loading arms.

Q5   Explain what the process of 'vessel certification' involves.

Q5   Answer

All vessels are required to be registered under a 'flag state' when they are first built, which means the owners of the vessel agree to comply with maritime regulations of the flag state. During the initial registration process, the vessel is inspected by the flag state inspection team and, provided it meets the required standards, it will be issued with a number of certificates, each relating to various aspects of the vessel.

Q6   Outline the responsibilities of the master of a seagoing vessel.

Q6   Answer

On any seagoing vessel, the person responsible for the safety of the vessel and all those aboard is the master. The master has the absolute right – and duty – to make the final decision on matters affecting the vessel and those on board.

Q7   Explain the difference between gangway boarding of a vessel and accommodation ladder boarding.

Q7   Answer

Gangway boarding is the main method of boarding a vessel. The gangway runs directly from the berth to the vessel at right angles to the vessel.

An accommodation ladder is an access ladder which is a permanent feature of the ship and is connected and is parallel to a ship's side. Its elevation can be adjusted according to the requirements of those wishing to board or leave the vessel.

Q8   Explain what a 'diving project plan' is.

Q8   Answer

A diving project plan will show how, and to what extent, the work to be done underwater will be split up into separate dive operations. It will also show how the hazards identified in the risk assessment will be controlled and it will include emergency and contingency plans.

274

Revision questions and suggested outline answers for Sub-element 5.2

Q1 Explain what purpose the UN classification and labelling system for the transportation of dangerous goods serves.

Q1 Answer

The system uses a standard signboard fixed to the vehicle in a designated position or positions. On this signboard are a number of internationally accepted series of codes and symbols to show what is being transported, the hazards that the cargo represents, and the correct preventative actions to be taken when required.

Q2 Outline FIVE measures which can be applied to avoid plant being struck by vehicles.

Q2 Answer – five of the following

- Traffic routes should be wide enough for the safe movement of the largest vehicle permitted to use them.
- Traffic routes should have enough height clearance for the tallest vehicle permitted to use them.
- Potentially dangerous obstructions need to be protected.
- Traffic routes should be planned to give the safest routes between calling places.
- Routes should avoid passing close to such things as unprotected fuel or chemical tanks or pipelines.
- All potentially vulnerable plant should be protected from errant vehicles.
- All routes should be well lit.
- Any hazardous sections of the route should be clearly signed.
- A person should be appointed to be responsible for, and oversee, site traffic movements on site.
- Drivers should be trained and authorized to drive vehicles on site.
- Visiting drivers should be briefed on site traffic movement rules.
- Reversing of vehicles should be avoided or controlled.

Q3 **Explain** the purpose of driver training for the transportation of dangerous goods.

Q3 Answer

The main objectives of driver training, in relation to carrying dangerous goods, are to ensure drivers are aware of the hazards arising when they are driving a vehicle which is carrying dangerous goods. Drivers should know what steps to take in order to reduce the likelihood of an incident taking place. Drivers should know what necessary measures they need to take when driving a vehicle which is carrying dangerous goods with regard to ensuring their own safety, that of the public and the environment. Also, if an incident does occur, drivers should know how to limit the effects of that incident. Drivers should have practical experience of what actions they will need to take in the case of an incident occurring.

Q4 **Outline SIX** control measures which can be implemented when the loading and unloading of hydrocarbon materials takes place.

Q4 Answer – six of the following

▷ The area designated for loading and unloading should be situated away from general traffic routes. It should also be situated on level ground.

▷ There should be sufficient space to allow the largest planned-for vehicles to easily manoeuvre into and out of the loading/unloading area.

▷ Loading/unloading areas should be adequately lit when in use.

▷ A system should be implemented that ensures a vehicle cannot be driven away from the loading/unloading point before being authorized to do so.

▷ When the loading/unloading operation has been completed, drivers must ensure all tank openings, including valves and caps, are closed before starting their journey.

▷ No tank should be overfilled, as most tanks require room for expansion of the liquid.

▷ When filling tanks, the pressure in the tank must be monitored to ensure it does not exceed its maximum working pressure. Tanks should be fitted with pressure relief valves.

276

- When discharging tanks, the pressure in the tank must be monitored to ensure a vacuum is not created. Tanks should be fitted with vacuum breaker valves.
- All external vents should be fitted with flame arrestors.
- The rate of filling or discharge must be limited to reduce the risk of static electricity building up.
- Suitable drip trays should be placed beneath hose connection points when loading/unloading operations are being set up or are being ended.
- Where product with a flash point of 60°C or less is being loaded/unloaded, a bonding wire should be connected between the vehicle and an earthing point before loading/unloading commences.
- A no smoking policy should be established and maintained on site.
- There should be two opposing emergency exits from the loading/ unloading area.
- Exits should be clearly marked and open outwards.
- Vapours which are displaced during the transfer operation should be returned to the donor tank via a vapour-tight connection line.
- The vapour return line should have a different connection fitting compared with the product transfer hose to ensure there can be no misconnection.
- The vapour return line should be connected before the product transfer hose is connected.
- There should be a device on the vehicle which locks the brakes in the 'on' position when the vapour recovery line is connected.
- A competent person should be given the responsibility to monitor all the hose connections during loading/unloading operations.
- Any uncontrolled release of vapour should be recorded in the vehicle log book and reported to the authorities.
- There should be a pre-formulated spillage plan ready to deal with any spillages, and a spillage kit at the ready. This kit should include bunding.

Q5 Give FOUR control measures which can be implemented to manage vehicle movements safely on site.

Q5 Answer – four of the following

▷ A person should be appointed to be responsible for, and oversee site traffic movements on site.

▷ All routes should be well lit.

▷ Any hazardous sections of the route, such as sharp bends or adverse cambers, should be clearly signed.

▷ Drivers should be trained and authorized to drive vehicles on site.

▷ Visiting drivers should be briefed on site traffic movement rules.

▷ The reversing of vehicles should be avoided or controlled.

▷ A contingency plan should be in place, along with the resources to implement it in case of adverse weather conditions such as snow and ice.

▷ Periodic surveys and safety tours should be conducted to ensure traffic rules are being complied with.

Q6 Give FIVE examples of checks which should be carried out prior to a vehicle carrying hydrocarbon material leaving site.

Q6 Answer – five of the following

▷ All hoses are secure.

▷ There are no leaks from the hoses.

▷ Blanking caps are fitted.

▷ The load is not leaking.

▷ The load is not overheating.

▷ For liquefied petroleum gas tanks, the pressure is within prescribed limits.

▷ The brakes and tyres on the vehicle are in good working order.

▷ All documents are in order and available.

▷ The markings are correct and in place.

Q7 Outline FOUR control measures which are specifically applicable to the loading/unloading of hydrocarbon materials from rail vehicles.

Q7 Answer – four of the following

▷ During loading/unloading operations, warning signs should be displayed on the train if open to access. These will include a red flag during the day, a red light at night and a warning sign that the rail cars are connected.

- Loading/unloading operations must be monitored throughout the operation.
- Rail cars which have been disconnected from the locomotive must be prevented from moving.
- There should be a 15-metre exclusion zone around any loading/unloading point.
- Where tools have to be used, these should be non-sparking tools.
- Prior to loading/unloading operations commencing, a system of vapour control should be established.
- The closure of all foot valves, lids and the removal of hoses, etc. must be overseen by a competent person.
- If rail tankers are fitted with product heaters, these must *not* be used if flammable vapours are present during loading and discharge operations.
- When work involves working on top of rail cars, working-at-height regulations should be adhered to.
- Weather conditions (the potential for lightning) should be taken into consideration prior to the commencement of loading/unloading.

Index

ABBI (look Above, Below, Behind, and Inside) technique 7
absorption removal 91
accident, definition 4
accident records, previous 8
accident/incident report forms 76
accommodation ladder 203
action plan formulation and implementation 7
active fire protection systems: chemical fire extinguishing systems 172; foam based extinguishing systems 170; inert based extinguishing systems 173; sprinkler systems 170; water deluge systems 170
additives in controlling electrostatic charges 145
Advanced Medical Aid (AMA) training 184
air stripping 91
air-sea rescue 190
alarm zones 168
alarms 180; detection zones 167; fire compartments 167; offshore 184; onshore 183–4; responses to 180; systems 167; tankers 134; verification of 92
analysis of information 7
ancillary equipment, inspection and assessment of 92
annular rim, failure of (bottom rim of storage tank) 102–3
anti-foaming agents 17
anti-wetting agents 17
antifreeze 90
Aqueous Film Forming Foam (AFFF) 171
argon (Ar) in fire extinguishing systems 173
arson 137
As Low As Reasonably Practicable (ALARP) 24, 25, 31
asphyxiation 16, 17

aspiration smoke detectors 167
asset integrity 80
atmospheric drain 119
auto-ignition temperatures 84

barrier cream 33
barrier modelling 33
barriers: control measures 33–4; effective 34
Basic Offshore Safety Induction and Emergency Training (BOSIET) 188–9
bentonite 20
Billy Pugh 203, 204
blanket gas 85
blow down (vent) valves and system 114; boiler 158; facilities 118–19
Blowout Preventers (B-O-Ps) 80
boiler/furnace components 154
boiler/furnace hazards 156–8; firebox overpressure 157; flame impingement 157; low tube flow 157–8; over-firing 156; pilot lights 156; Tube Metal Temperature (TMT) 158
boiler/furnace maintenance 158–9; boiler blow down 158; logs and boiler/furnace checklists 159
boiler/furnace safety components 154–5
boiler/furnace, use of 154–6
boilers: definition 154; types 155
Boiling Liquid Expanding Vapour Explosion (BLEVE) 16, 60, 134, 136, 143
bonding in controlling electrostatic charges 145–6
breaking containment permit 69
Breathing Apparatus (BA) 187
brittle fracture 101; in storage tanks 127
brittleness 99
Buncefield incident (2005) 31
bunding 133
burns 17, 18, 143
butane 16

Index

calibration of vessels and instruments 92
camera-based flame detectors 164
carbon dioxide (CO_2): in fire extinguishing systems 173; storage 135
carbon steel tanks 127
carcinogenic properties 15
causes of accidents and incidents 8–9; immediate causes 8, 9; root causes 8, 9; underlying causes 8, 9
CCTV coverage 8
cellulosic fire 169
Central Control Room (CCR) 196
centrifuge (spin the oil clean) 90
certification of marine vessels 200
check sheets 8
checklists 28
chemical attack in storage tanks 126–7
chemical fire extinguishing systems 172
cigarettes, as source of ignition 83
client (employing company) 48; responsibility 48
closed drain systems 119–20
coastguard agency 188
cold burns 17
cold weather clothing 204
commissioning 91–2
competency 81
compression 98
Computational Fluid Dynamics (CFD) models 35
condition monitoring maintenance 82
conductivity meter 158
configuration check, system 92
confined spaces entry certificate 69
Confined Vapour Cloud Explosion (CVCE) 136, 143
contractors 48–51; instruction and training 50; managing/supervising 49–50; permit-to-work systems and 69; responsibilities 50; risk assessment 50; safe handover procedure 51; safety 50; scale of use 48
control measures 21, 33–4, 55
control room design 60–1
control room operators (CROs) 164
control room structure 60–1
Convention Concerning International Carriage by Rail (RID) 213
cooling effect of foam 171
corrosion: preventatives 18; prevention 81; of steel 18; in storage tanks 125–6
covering effect of foam 171

creep 99; thermoplastic tanks and 126
cutting, as source of ignition 83

Dangerous Occurrence 4
Damage Only 4
decommissioning and dismantlement 42, 137–8
describe (command word) xvi
direct-fired heaters 84
direct-fired space 84
diver operations 204–6
diving certificate 69
diving contractor's responsibilities 205
diving project plan 205
documented evidence: content of 41–3; purpose of 40
domino effects 58
Dow/Mond indices 57–8
drainage 119–20; closed 119–20; interceptors 120; open systems 119
drained water 85
drill string 19
drilling muds (drilling fluid) 19–20
drills 189
driver training for transportation of dangerous goods 209–10
dryers, as source of ignition 83
ductility 98

elasticity 99
electric motors, as source of ignition 83
electric sparks, as source of ignition 83
electrical equipment, ignition source 147
electrical work permit 69
electromagnetic radiation 83, 147
electrostatic charges 143–6; controlling 145–6; controlling additives 145; controlling bonding and grounding 145–6; filling operations 144; filtration 145; other issues 145; piping systems 144; sparks 83, 147; water and 18
Emergency Breathing Systems (EBS) 186
Emergency Command and Control (ECC) centre 32, 182
Emergency Life Support Apparatus (ELSA) 186, 204
emergency plan 178–82; alarm systems and responses to alarms 180; contents 179–80; fire and explosion strategy 181–2; off-site emergency plan, triggering 180; off-site communication measures 180; on-site

178–9; responsibilities and authority of those overseeing 178; training and instructions 180; types of events planned for and extent of responses planned 179–80

Emergency Shutdown (ESD) system 112–16, 170, 182, 203; bypassing 117–18; components of 113–14; level of shutdown 114–16; logging actions and displaying warnings 118; typical actions 112; voting systems 113

Emergency Shutdown Valve (ESDV) 81, 113, 182

empirical models 35

environmental conditions at the time of accident 8

erosion in storage tanks 126

escape and evacuation: offshore 185–7; onshore 185

ethane 16

evacuation *see* escape and evacuation

exclusion zone, marine 196

expansion bellows 102

experiences 8

explain (command word) xvi

explosion consequence assessment modelling 35–6

explosion hazard assessment modelling 35; Computational Fluid Dynamics (CFD) models 35; empirical models 35; phenomenological models 35

explosions 143; leak and fire detection systems 164–7; plant layout and 58

external support agencies 189–90; liaison with emergency services 190–1

failure case selection 28

Failure Mode and Effects Analysis (FMEA) 27

Failure Modes and Effects and Criticality Analysis (FMECA) 29–31

failure modes 98, 99–101; brittle fracture 101; creep 99; failure of annular rim (bottom rim of storage tank) 102–3; introduction 98; knowledge of 101–2; stress 99; stress corrosion cracking (SCC) 100; thermal shock 100

Fast Rescue Craft (FRC) 187

fatigue in storage tanks 126

ferrous sulphide 84, 148

film formation effect of foam 171

filters 85

fire, plant layout and 58

fire, types of 169; cellulosic 169; hydrocarbon 169

Fire and Gas (F&G) detection system 164, 182

fire ball 35, 60

fire compartments 167

fire deluge systems 81, 148

fire detection systems 81, 114, 164–8

fire protection systems for spherical storage tanks 173–4; fixed monitor system 174; water deluge system 174; water spray system 175

fire teams: onshore 187; offshore 187–8

fire triangle 14, 15, 142

fire tube boilers 154

fire walls, coding of 169

firebox overpressure 157

fixed dry chemical powder fire extinguishing systems on bulk gas carrying ships 172

fixed roof storage tanks 132

flag states 200

flame detection systems 164–5; camera-based 164; combined ultraviolet/infrared 165; ultraviolet 165

flame impingement 157

flame retardant gloves 186

flammability 14; extremely flammable 14; flammable 14; highly flammable 14

flammable range 15; lower flammable limit 15; upper flammable limit 15

flare types 118–19; air assisted flares 118–19; multi-point pressure assisted flares 119; steam assisted flares 118; unassisted flares 119

flash fires (fire ball) 35, 60

flash point 14

floating assets 198

Floating Production, Storage and Offloading (FPSO) unit 198, 202

floating roof tanks 129–31, 173–4; external 129; internal 130; issues with 130–1; landing the roof 130; rim seal failures 131

foam based extinguishing systems 170–1; controlling effects of foam 171; how foam is made 170; polymer film additives 172

freezing water 18

friction heating 83

Frog transfer system 204

frostbite injury 18

furnace and boiler operations *see under* boiler/furnace

furnace types 155–6; natural draught 155; balanced draught 156
furnaces (process heaters) 83; definition 154; as source of ignition 83; types 155–6

gangway boarding 202–3
gas blow by 120
gas detection systems 81, 114, 165–6; infrared absorption combustible 165; open path infrared 166; point infrared 166
gas freeing and purging operations 85
gas turbine systems and compressor systems 175
General Platform Alarm (GPA) 183
genotoxic carcinogens 15
give (command word) xvi
Glass Reinforced Plastic (GRP) tanks 126, 127
gravity drain 119
gravity separation 90
grounding in controlling electrostatic charges 145–6

handback 67
handover 67
hazard, definition 4, 16
Hazard and Operability Study (HAZOP) 27, 28–9, 117; study flow chart 30; team members 29
hazard assessment 28
hazard checklists 28
Hazard Identification Study (HAZID) 27–8, 117
hazard realization 33
hazard signboard, tanker 208–9
hazardous area classification study 148
hazardous materials transportation, by road tanker 208–9
Health and Safety Executive (HSE) (UK) 32; HSG250 68
health and safety information 8
heat detection systems 165; linear heat detectors 165; point heat detectors 165
heating the oil dry 91
helicopter rescue 184, 187, 188
management of major incident risks 31–2
hierarchy of risk control 9, 25
High-Density Polyethylene (HDPE) tanks 126

High Integrity Pipeline Protection Systems (HIPPS) 113, 182; valves 113
historical data analysis 27
hook up 91–2
hot work permit 66
hydrocarbon fire 169
hydrocarbon vapour clouds 136
hydrogen 15
hydrogen sulphide (H_2S) 16

identify (command word) xvi
ignition, sources of 83–4, 146–8
illegal tapping 137
impact sparks 83
industry related process standards 32
inert based extinguishing systems 173
inert gas (blanket gas) 85; systems 173
inerting operations 85
information 8–9
infrared absorption combustible gas detectors 165
inherent safe and risk based design concepts 32–3, 42
injury 4; major injury 4; minor injury 4; significant injury 4
inlet separators 85
instrument readouts and records 8
insulating effect of foam 171
interceptors 120
International Maritime Organization (IMO) 200; International Code for the Construction and Equipment of Ships Carrying Liquefied Gases in Bulk (IGC Code) 172; Safety of Life at Sea (SOLAS) regulation II-2/1.6.2 172
interview techniques 7–8
investigations of accidents and incidents: analysis of causes 9; benefits from 5–6; financial reasons for 5; legal reasons for 4–5; reasons for 4–5; four stages process 7; team make-up 6; training for investigating team 6
ionization point smoke detectors 166

jet (spray) fires 35, 60, 135

land transport 207–15
landing the roof 130
Lead-210 20
leak detection systems 167
learning from incidents 9–10
Legionella 18
leptospira 18

ife rafts 187
ifeboat rescue 186–7, 190
ifejackets 204
ightning 83, 142, 147; power surges
    142; sparks 142
inear heat detectors 165
Liquefied Natural Gas (LNG) 16–17, 135
Liquefied Petroleum Gas (LPG) 16, 134
oading and discharging arrangements,
    road tankers 210–11
ock out device 70
ock out, tag out (LOTO) and isolation 70
ogs 8
ow specific activity sludges 20–1;
    control measures 21; hazards 21
ow tube flow 157–8
Lower Explosive Limit (LEL) 157

Magnetic Particle Inspection (MPI) 107
Major Accident Prevention Policy
    (MAPP) 43, 179
malleablility 98
Man Overboard (MOB) 197
management of change 61
management of simultaneous
    operations (SIMOPS) 138
manufacturers' instructions 8
marine transport 195–206; certification
    of vessels 200; diver operations
    204–6; exclusion zone 196; loading
    and unloading at marine terminals
    198–200; Personal Protective
    Equipment (PPE) for 204; personnel
    transfers and boarding arrangements
    202–4; roles and responsibilities of
    marine co-ordinator 202; roles and
    responsibilities of masters and crew
    200–2; transfer of material between
    marine vessels and tanks 198–200;
    vessel hazards 197; vessel types
    197–8
matches, as source of ignition 84
Material Safety Data Sheet (MSDS) 17,
    18
Maximum Continuous Rating (MCR),
    156
measurements 8
medical emergency planning 183–4
medical response 183–4; offshore 184;
    onshore 183–4; triage 184
medevac (medical evacuation) 184
mercaptans 19
methane 16
method statements 8

methyl mercaptan 19
micro-biocides 18
mist nozzles 175
monoammonium phosphate based
    chemicals 172
Multi-Point Ground Flares (MPGFs) 119

Naturally Occurring Radioactive
    Materials (NORM) 20
near misses 4, 6
nitrogen 17; in fire extinguishing
    systems 173
Non-Destructive Testing (NDT) 107
non-genotoxic carcinogens 15
non-hazardous place 148

observational techniques 7–8
Occupational Health and Safety
    Administration (OSHA) 15, 32
offshore installations 40
offshore safety case and safety
    reports 41–2; availability 42;
    combined operations 42; control
    of major accident hazards 41;
    decommissioning and dismantlement
    42; factual information 41; life cycle
    requirements 42; major accident
    hazard identification 41; major
    accident risk evaluation 41; major
    accident risk management 42;
    management of health and safety 41;
    rescue and recovery 42; safe design
    concept 42; Safety Management
    System (SMS) 42
oil based drilling mud 20
oil loading arms 200
onshore installations 40
onshore safety case and safety reports
    40, 42–3; descriptive information
    42; incident prevention information
    43; offsite emergency plan 43;
    emergency response measures
    43; potential major incidents 43;
    prevention/mitigation of major incident
    43; onsite emergency plan 43
open drain systems 119
open path infrared combustible gas
    detectors 166
opinions 8
optical beam smoke detectors 166
optical point smoke detectors 166
orinasal masks 204
outcome of accidents and incidents 4
outline (command word) xvi

overfilling 133–4
over-firing 156
oxygen 17

passive fire protection 168–9; types 168
permit-to-work system 21, 50, 84, 118;
    appropriate use of 65–6; authorization
    and supervision 67; display 66;
    handback 67; handover 67; interfaces
    with adjacent plant 69; interfaces with
    contractors 69; key features 64–5; lock
    out, tag out (LOTO) and isolation 70;
    objectives and functions of 65; permit
    interaction 67; records 8; role and
    purpose 64; suspension 66; template
    68
permits, types 68–9
Personal Protective Equipment (PPE) 9,
    25, 50; for marine use 204
Personnel On Board (POB) 183, 203
phenomenological models 35
photographs 8
pilot ladder boarding 203
pilot lights 147, 156
Pipeline Inspection Gauges (PIGS) 137
pipelines, protection 137
plans and diagrams 8
plant layout 57–60; aggregation/trapping
    of flammable vapours 60; control
    room design 60–1; control room
    structure 60–1; domino effects 58;
    Dow/Mond indices 57–8; explosion
    58; fire 58; inherent safety 57;
    positioning of occupied buildings
    59–60; reduction of consequences of
    event on- and offshore 59; Temporary
    Refuge (TR) integrity 61; toxic gas
    releases 59
platform support vessel (PSV) 197
point heat detectors 165
point infrared combustible gas detectors
    165
point smoke detectors 166
pollution 197
polymer film additives 172
pool fires 34, 60, 135
pre-start-up safety review 82–3
Pressure and Vacuum relief valves
    (P&Vs) 132
pressure burst in control room 60
pressure drain system (closed drain
    system) 119–20
prevention measures 43
process designers 32

process drawings 8
Process Hazard Analysis 54
process heaters see furnaces
Process Safety Management (PSM)
    54–7; contractors 56; emergency
    planning and response 57; employee
    participation 55; health and
    safety considerations 55; incident
    investigation 57; management of
    change 56; mechanical integrity
    56; operating limits 55; operating
    procedures 54–5; permits-to-work 56;
    pre-start-up safety review 56; process
    hazard analysis and 54; training 55
production separators 85
propane 16
properties of materials 98–9
Protective Systems Isolation Certificate
    (PSIC) 118
pyrophoric scale 147

qualitative risk assessment 25
quantitative risk assessment 27

radiation 5, 21; electromagnetic 147
radiography testing 108
radium 20
rail tankers 213–14; and tanks, transfer
    of material between 133–4
records and sources of information 8
refrigerants 18
Remotely Operated Shut-Off Valves
    (ROSOVs) 113
residual risk 4
Respiratory Protective Equipment (RPE)
    21
rim seal fire-fighting unit 173
risk, definition 4
risk assessment 8; definition 24–5;
    five-step approach 25; historical
    data analysis 27; purpose of 24–5;
    qualitative risk assessment 25;
    quantitative risk assessment 27;
    risk rating/prioritization 26; semi-
    quantitative risk assessment 25–6
risk based maintenance 82; inspection
    82
risk control measure 9
risk rating matrix 26
risk rating/prioritization 26
road tankers 208–11; and tanks, transfer
    of material between 133–4
running fire 34
rusting 17

safe operating envelope 101
safety case and safety reports 40; offshore 40; onshore 40
safety critical elements (SCEs) 80–1; blowout preventers 80; corrosion prevention 81; Emergency Shutdown Valve (ESDV) 81; fire and gas detection systems 81; fire deluge systems 81; safety inspection and testing 81; training and competency 81
safety inspection and testing 81
Safety Instrumented Systems (SIS) 116
Safety Integrity Level (SIL) 116–17
Safety Management System (SMS) 42
safety reports, definition 40
scramble net 186
sea ladder 186
sea water 19
Search and Rescue (SAR) helicopters 184
semi-quantitative risk assessment 25–6
separating effect of foam 171
shear 98
shift handover: 12-hour 74; background 74; importance of 75; key principles 75–6; plant and equipment operation and 76–7; two-way communication/ joint responsibility 74–5
shutdown protocol 92
single buoy mooring (SBM) 198
Single Point Moorings (SPM) 198
sketches 8
skin irritant 15
skyscape 186
slug catchers 85
SMART objectives 9
smoke detection systems 148, 166–7
smoke hoods 186
smoking 146
sparks 83; electrostatic discharge 147; from lights and switches 147
spot fire detection systems 164
sprinkler systems 170
stainless steel tanks 127
Standby Vessel (SBV) 184, 188, 196
start-up and shutdown: operating instructions and procedures 88–9; safe 88–9; shutdown 89; thermal shock 89
steam 19
steel, corrosion of 18

storage tanks 124–5; alarms 134; bunding 133; capacity 124; evolving damage mechanisms 125–7; excessive external loads 128; filling 133–4; fixed roof storage tanks and vacuum hazards 132; floating roof tanks 129–31; impact 127; integrity management 124–5; mechanical damage 127–9; overfilling 133–4; over-pressurization 128; pressurized/ refrigerated vessels 134–5; settlement which is non-uniform 127–8; tanks floating off their foundations 128–9; transfer of product 133–4; vacuum 128; wind loads 128; loss of containment and consequences 135–7; pipelines, protection, surveying, maintenance, security against arson and illegal tapping 137
strategic command posts 189–90
stress 99
stress corrosion cracking (SCC) 100
sub-contractors 49
Subsea Isolation Valves (SSIVs) 182
suppression effect of foam 171
surge drums 85
switches, as source of ignition 83
synthetic based drilling mud 20

tag out device 70
technical information 8
temperature classification in zoned areas 149, 150
Temporary Refuge (TR) 61, 182, 186
tension (tensile force) 98
thermal radiation 143
thermal shock 100
thermocouple 156
thorium 20
three-phase separators 85
toolbox talk sheets 8
torches 186
Total Dissolved Solids (TDS) meter 158
toxic gas release in control room 60
toxicity 15; acute 15; chronic 15
traffic management 212; vehicle movements offsite 212; vehicle movements onsite 212
training 31, 55, 81; driver, for transportation of dangerous goods 209–10; for emergency plan 180; for investigating team 6; for offshore installations 188–9

training records 8
triage 184
Tube Metal Temperature (TMT) 158
tundishes 119

ullage 129
ultrasonic flaw detection 108
ultraviolet flame detectors 165
ultraviolet/infrared flame detectors 165
Unconfined Vapour Cloud Explosions (UVCEs) 136, 143
uranium 20

vacuum dehydration 91
vapour cloud explosion (VCE) in control room 60
vapour cloud fire (VCF) 35
vapour density 14
vapour pressure 14
vehicles, as ignition source 84, 146
vent system 114
venting operations 85
Very Early Smoke Detection Apparatus (VESDA) 182
vibration analysis 82
victim statements 8
voting systems 113

walking the line 92
water 18–19; controls 90; freezing 19; presence and removal 90; removal of 90
water based drilling mud 20
water deluge systems 170
water mist systems 173
water tube boilers 155
weld failures 106–8; cold cracks 106; cracks 106; hot cracks 106; introduction to 106; porosity 107; types 106
weld profile 107
weld testing techniques 107–8
welding, as source of ignition 83
wet weather clothing 204
Widdy Island Disaster (1979) 199
witness statements 8
work permit see permit-to-work
working at height permit 69
Worldwide Offshore Accident Databank (WOAD) 27
written procedures 8

Yokohama fenders 198

zoning 148–9; selection of equipment 149–50